UNDER THE SEA-WIND

RACHEL CARSON (1907–1964) is widely acknowledged to be one of the most influential writers of the twentieth century. *Under the Sea-Wind* (1941) was the first of her distinguished trilogy on the sea. It was followed by *The Sea Around Us* (1951) and *The Edge of the Sea* (1955), and catapulted Carson to international fame as a nature writer. Carson's prophetic warning about the misuse of chemical pesticides, *Silent Spring* (1962), is considered the touchstone of the contemporary environmental movement.

LINDA LEAR is an environmental historian and biographer. The author of the prize-winning *Rachel Carson: Witness for Nature* and of *Beatrix Potter: A Life in Nature,* she has also edited *Lost Woods: The Discovered Writing of Rachel Carson* and frequently writes and speaks about Carson and her legacy.

RACHEL CARSON

Under the Sea-Wind

Introduction by
LINDA LEAR

Illustrations by
HOWARD FRECH

PENGUIN BOOKS

PENGUIN BOOKS
Published by the Penguin Group
Penguin Group (USA) Inc., 375 Hudson Street, New York, New York 10014, U.S.A.
Penguin Group (Canada), 90 Eglinton Avenue East, Suite 700, Toronto,
Ontario, Canada M4P 2Y3 (a division of Pearson Penguin Canada Inc.)
Penguin Books Ltd, 80 Strand, London WC2R 0RL, England
Penguin Ireland, 25 St Stephen's Green, Dublin 2, Ireland (a division of Penguin Books Ltd)
Penguin Group (Australia), 250 Camberwell Road, Camberwell,
Victoria 3124, Australia (a division of Pearson Australia Group Pty Ltd)
Penguin Books India Pvt Ltd, 11 Community Centre, Panchsheel Park,
New Delhi – 110 017, India Penguin Group (NZ), 67 Apollo Drive, Rosedale, North Shore,
Auckland 0745, New Zealand (a division of Pearson New Zealand Ltd)
Penguin Books (South Africa) (Pty) Ltd, 24 Sturdee Avenue,
Rosebank, Johannesburg 2196, South Africa

Penguin Books Ltd, Registered Offices:
80 Strand, London WC2R 0RL, England

First published in the United States of America by Simon and Schuster 1941
Published in Penguin Books 1996
This edition with an introduction by Linda Lear and a new selection of illustrations published 2007

9 10 8

Illustrations by Howard Frech
Pages 2, 49, 85, 99, 103: Courtesy of the Lear/Carson Collection, Connecticut College
Pages 4, 28, 141: From the collection of the Ward Museum of Wildfowl Art,
Salisbury University, Salisbury, Maryland

LIBRARY OF CONGRESS CATALOGING IN PUBLICATION DATA
Carson, Rachel, 1907–1964.
Under the sea-wind / Rachel L. Carson.
p. cm.—(Penguin classics)
Includes bibliographical references.
ISBN 978-0-14-310496-4
1. Marine biology—Atlantic Ocean. I. Title.
QH92.C3 2007
578.77'3—dc22 2006050707

Printed in the United States of America
Set in Sabon

To my mother

Contents

Introduction

Under the Sea-Wind, Rachel Carson's literary debut, evolved inauspiciously from an eleven-page introduction to a government fisheries brochure. It marked the debut of one of the finest writers of English in the twentieth century, and a scientist who ultimately changed the way we view our relationship with nature. Although critically acclaimed, her first effort was denied broad appreciation by the outbreak of World War II, which greeted its publication in 1941. The disappointed author bought back the unsold copies and gave them away to treasured friends. But she never gave up on the idea that it would one day be republished. *Under the Sea-Wind* was inspired quite simply by Carson's love of the mysterious and the wonderful. It remained her favorite of all of her books, recalling a rare, peaceful time in her life.

In April 1936, Carson was an unemployed zoologist and freelance writer for the U.S. Bureau of Fisheries assigned to write radio scripts on ocean life. By night she earned money writing articles on the natural history of the Chesapeake Bay for *The Baltimore Sun,* signing them "R. L. Carson" in an effort to convince her readers that she was male, and thus take her science seriously. Her bureau chief at Fisheries, Elmer Higgins, had assigned her the job of writing a general introduction to marine life for a new brochure. Her mother, Maria, neatly typed her essay, "The World of Waters," in the small elite typeface of the old Smith Corona typewriter, and the next day Carson sat in Higgins's Washington, D.C., office waiting for his verdict. The government ichthyologist knew at once that it was unsuitable. What he was reading was a piece of literature. Carson never forgot the conversation: "My chief . . . handed it back with a twinkle in his eye. 'I don't think it will do,' he said. 'Better try again. But send this one to the *Atlantic [Monthly]*.'" Pleased by his response, Carson quietly put her essay away in a desk drawer and went on with her assignment. Almost a year later, economic reality impelled her to try to sell it to the *Atlantic Monthly*, the top literary magazine of the day, as Higgins had suggested.

Reading even the earliest drafts, it is easy to understand Higgins's enthusiasm, and to see why the *Atlantic Monthly* bought and published Carson's piece in September 1937, retitled simply "Undersea." Her essay was a narrative account of the myriad creatures of the undersea world, and introduced two of Carson's most enduring literary themes: the ecological relationships of ocean life that have endured for eons of time, and the material immortality that embraces even the tiniest organism. In it she conducts the reader on a tour of the deepest ocean floor from the immediate point of view of an underwater eye, describing each scene scientifically yet with such wonder and delight that the underwater world is made accessible to the most nonscientific reader. "Undersea" introduced Carson as a writer of critical interest, and established her unique voice, at once scientifically accurate, yet poetic and lyrical. In the ecology of the sea, Carson had not only found something she loved to write about but also the medium through which she could share her vision of nature's oneness. From these four pages in the *Atlantic [Monthly]*, Carson later admitted, "everything else followed."

The young zoologist who wrote "Undersea" had been a keen observer of the natural world, particularly of birds, since her western Pennsylvania childhood. She had always loved to write, and nature had been what she knew and delighted in most. In college Carson took the unusual step of changing her major from English to zoology under the influence of a dynamic female professor. Scholarships to the Marine Biology Laboratory at Woods Hole in Massachusetts in 1929 and to Johns Hopkins University in Baltimore, Maryland, for graduate work established her scientific credentials. But there were no jobs for women in science in the Depression years of the 1930s. Carson's nearly perfect scores on the civil service examinations allowed Higgins to offer her a federal appointment in 1937, but it was her obvious literary skills that moved her up the ranks of the U.S. Fish and Wildlife Service for the next fifteen years.

Like most ambitious writers in their novitiate, Carson needed a literary mentor to move ahead. Quincy Howe, then senior editor at Simon & Schuster, and one of his best authors, the renowned journalist, cultural historian, explorer, and illustrator Hendrik van Loon, had been impressed by "Undersea." Van Loon, whose initial letter to Carson memorably arrived in an envelope splashed with the green waves of a sea through which sharks and whales poked inquisitive snouts, wanted to know what else Carson knew about what went on under the sea. It was that question that made her think seriously about writing a book. In January 1938, Van Loon arranged a meeting with Howe, at which Carson presented an outline for a book of a dozen or so chapters introduced by a preface based upon her earlier essay.

She initially thought of a narrative account of the daily life of several sea creatures, much in the manner of the great English naturalist Henry Williamson, whose work she deeply admired. She would divide the book into three parts, one for the life of the shore, "The Edge of the Sea," one for the open sea, "The Gull's Way," and one for the deep abyss, "River and Sea." In each of these parts, she would tell the story of one particular animal: a gull, a mackerel, and an eel. Taken together, the three narratives would weave a tapestry in which the ecology of the ocean and the interdependence of all its creatures would emerge. The central character, however, was the sea itself. To be successful, Carson needed imagination, acute observation, and a comprehensive scientific understanding of the ocean and its inhabitants.

Even more than the financial help that she sorely needed, Carson wanted to go "undersea" herself. She appealed to Van Loon to introduce her to his friend William Beebe, the distinguished oceanographer and ornithologist who was, at the time, director of tropical research for the New York Zoological Society. Carson explained that a dive would give her "the feeling of the water as no amount of vicarious experience could do." Sadly, her first and only underwater dive, when it did come a decade later, was in less-than-ideal conditions, but for Carson even a murky experience was transforming.

As a fisheries' biologist, the breadth of Carson's understanding of natural processes was not unusual, but her response to the natural world, and her sense of wonder and delight in it, distinguished her attitude toward her research as well as her prose. Carson went first to Beaufort, North Carolina in the summer of 1938, taking her mother and the two young nieces whom she supported with her. The fisheries station at Beaufort was then the largest research facility on the East Coast next to the one at Woods Hole, and had the advantage of a wide expanse of ocean beach. Carson sought out the most remote sections, finding a particularly lovely wild stretch that she used as the background for the chapters about shore birds. She had timed her visit to Beaufort for the tides and full moon. No matter how hard the wind blew, she walked the beach at all hours. During the day she took notes on the comings and goings of shorebirds, observed the smaller shore creatures such as the crabs and sand fleas, and collected material. Sometimes she simply lay in sandy dunes, flat on her back, arms behind her head, watching and listening to the birds as they wheeled and screeched overhead.

She especially loved to wander the beach by night. It was the start of an adventure that became a lifelong habit. Flashlight in hand, she watched the nocturnal creatures come out of hidden homes, unseen by even the most observant eye during the daytime. Jotting down notes about her nighttime observations in a tiny black notebook, she included the sounds and smells of the night sea, the surf, the wind, and the pines on higher ground, and the stillness of the ponds at high tide. These images would give her writing some its most distinctive motifs.

Carson also discovered the marsh pools and ponds in the salt sands where the dunes of the barrier island fell away to the ocean. She sat for hours, totally enraptured, watching wave after wave pour through the slough into the ponds, where the high water released thousands of small fish that had been held captive since the previous spring. As she watched them race down the slough to the ocean, she was profoundly moved by the awe she felt at the mystery of life. Carson fell in love with the outer banks that summer, and with the mysterious relationship between sea and shore. No matter how many times she returned, the wonder of that place as she knew it then remained vivid.

The following summer, 1939, Carson headed north to the Woods Hole Station. There she worked in the library and the laboratory, but mostly walked the beaches along the harbor and sat on the fisheries wharf filling up her little notebooks with observations on the fish, and watching the tides bring their treasure in and out of the deep ocean channel. The experience gave her a greater understanding of the life of schooling fish and the ebb and flow of life "undersea."

Carson was a slow, painstaking writer. With her government job consuming her waking hours, she began the habit of working late in the evening or early in the morning when the house was quiet. Her creative process required real solitude, and she preferred silence, finding herself distracted by even the normal household noises. She revised paragraph by paragraph, sometimes even sentence by sentence, before going on to the next. Among the preserved manuscripts for *Under the Sea-Wind* are seven drafts of one page of the chapter "Spring Flight," each heavily corrected. Conscious of the role of alliteration and rhythm to create atmosphere, she read pages aloud to herself before she had her mother read them back to her. While Carson was at work, her mother typed the revisions so they would be ready for her daughter's nighttime labor. This was a pattern the two women held to with every piece Carson wrote, until her mother passed away in 1958. Each draft was read aloud, over and over, until Carson was satisfied with the way it sounded as well as the way it looked to the eye.

In the early spring of 1940, Carson sent Quincy Howe the five chapters that comprise Book I. Anxious months passed. It was not until June that Carson received a contract, an additional advance, and a definite deadline of December 1940. Assured of publication, she went into high gear, discovering that writing under pressure was not such a bad thing. The last week of July 1940, Carson returned to the Fisheries Station in Woods Hole to complete her research on the deep ocean floor.

As a full-fledged government employee, Carson had regular

access to sail on the *Phalanthrop,* the small fisheries dredger that daily steamed up and down Vineyard Sound or Buzzards Bay. When, as a graduate student, she had first beheld the treasures brought up from the ocean bottom, she had wondered where those creatures lived and how they survived. Now she had the scientific knowledge that allowed her imagination to go down through the water and see "the whole life of those creatures as they lived in that strange world."

Carson spent hours on the fisheries dock, just as she had on the beaches of Beaufort. She watched the schools of mackerel moving up and down along the stone breakwater, with squid and other predators darting among them. She must have made a pretty picture, sitting on a large wharf pile, a tiny woman of thirty-three, clad in a simple white blouse and cotton slacks, the wind blowing her light-brown hair around her oval face. She filled her notebooks not simply with lists of creatures, but once again with the sounds, colors, and motions of their existence.

Carson had definite ideas about how she wanted her book to look. She hired Howard Frech, an artist she knew and had confidence in on the staff of the *Baltimore Sun,* to make eight full-page pencil drawings and an illustrated glossary that would identify the more than one hundred sea creatures that populated her story. She convinced Simon & Schuster to accept a local artist, but she had to pay for his illustrations. The drawings were published with a deep blue wash, in keeping with the sense of mystery inspired by her text. The glossary gave the book the necessary scientific legitimacy, filled as it was with accurate as well as fascinating descriptive material Carson could not include in the text.

Her final trip to Woods Hole helped Carson make the necessary adjustments in her thinking, and most of all immerse herself in "a world that was entirely water." When she returned to her home in Silver Spring, Maryland, she had all she needed to finish the book. Carson's mother flawlessly typed the finished manuscript, and Carson sent it off on New Year's Eve 1940. Editors at Simon & Schuster still tell the story of how *Under the Sea-Wind* was the only manuscript ever received that had not a single typographical error. Years later Carson looked back at the writing of this book as one of the happiest times of her life. It had been pure joy to learn about the sea from bottom to top. She never recalled those months of 1940 without a pang of longing for that evanescent sense of total absorption in the creative process.

Under the Sea-Wind is an adventure in the ecology of air and water. While "Undersea" had given a submarine view of the ocean floor, *Under the Sea-Wind,* which originally had as its subtitle "A Naturalist's Picture of Ocean Life," was an intimate portrait of sea and shore creatures whose world of air and water the reader enters as an observer of events. Although the three parts of the book focus on different protagonists—sea birds, Scomber the mackerel, and Anguilla the eel—the whole is tied together by lives lived intimately with the sea. The words "sea wind" were Carson's shorthand for the encapsulation of all life within a single system. She subtly introduces us to the tie-lines of ecology in which each entity is linked and made whole by an integrated cycle of life. The book presents a new writer with a new way of thinking about the natural world, with which we are in an intimate partnership. As in the books that followed, *The Sea Around Us, The Edge of the Sea, The Sense of Wonder,* and *Silent Spring,* Carson takes us on a journey with the intent to change our attitudes about our relationship with the natural world.

The plot of *Under the Sea-Wind* is formed by each creature's struggle to survive and reproduce. It is not a story of Darwinian determinism molded by some fierce struggle, but by the role of chance. Survivors are those who are in the right place at the right time. Danger is everywhere, yet Carson's narrative of sea life conveys an overall sense of calm. Everything is as it should be in the pattern of an ancient, sometimes violent, but endless cycle comforting in the certainty of repetition. What marks Carson's writing is not her scientific interest in the impersonal forces of nature as they act upon her protagonists, but her sympathetic identification with individual creatures with whom she establishes a spiritual as well as a physical connection. In *Under the Sea-Wind,* Carson confronts one of the central problems of nature writing: how to give the processes of nature metaphorical and spiritual meaning without compromising the scientific accuracy of each creature's struggle for survival. Her voice is that of both scientist and poet, in love with the wonder she has discovered in nature.

Under the Sea-Wind was also a product of its time, reflecting anxiety for the future, yet offering the reassurance of nature's constant change. The world was at war, and death was all too close a reality to every man and nation. Carson's awareness of "material immortality" pervades her narrative, and speaks to the larger problems of human existence. The death of one creature contributes to the life of another in an endless chain of reincarnation. The sea-wind binds equally within its breath the lives lived in air and those in water, and, she infers, those on land. The ceaseless flow of life and death revealed under the sea-wind lent a certain optimism to the struggle of human existence everywhere.

Carson's prose, as in all her subsequent work, was influenced by the romantic school of nature writing exemplified by the Englishman Richard Jefferies. In *Under the Sea-Wind,* she draws particularly on his *The Pageant of Summer* (1905), in which he lyrically described how "the whole office of Matter is to feed life." His idea of the sea as both a real and spiritual source of life was essential to Carson's understanding of the natural world. In later years, she always kept a copy of Jefferies's poetry by her bedside. The title for her first book comes from one of her favorite passages in *The Pageant of Summer*: "As the wind, wandering over the sea, takes from each wave an invisible portion, and brings to those of shore the ethereal essence of ocean, so the air lingering among the woods and hedges—green waves and willows—full of fine atoms of summer."

In her effort to make the sea and its life a reality, Carson asks the reader not only to exercise his imagination of what it is to be a creature of the air and of the sea, but also to abandon the human yardstick of time. "Time measured by the clock or the calendar means nothing if you are a shore bird or a fish," Carson writes in her foreword, "but the succession of light and darkness and the ebb and flow of the tides mean the difference between the time to eat and the time to fast, between the time the enemy can find you easily and the time you are relatively safe. We cannot get the full flavor of marine life—cannot project ourselves vicariously into it—unless we make these adjustments in our thinking." It was a challenge that most readers were willing to accept. But a second suspension of rational thought was more difficult.

In order to make a fish, a shrimp, a comb jelly, or bird appear more real to human readers, Carson attributed to her creatures human traits and expressions that would not be acceptable in most scientific writing. She took the risk of near, but never total, anthropomorphism, certain in her own mind that by using terms that approximate human psychological states, the behavior of the creature would be understandable to her readers. "I have spoken," she writes, "of a fish 'fearing' his enemies, for example, not because I suppose a fish experiences fear in the same way that we do, but because I think he *behaves as though he were frightened.*" It was a literary risk, but in Carson's skilled hands there is no jarring juxtaposition from human to animal attribute. The leap is made seamlessly.

Under the Sea-Wind was published November 1, 1941, and sold for $3.00 a copy. The cover was a dull teal with beige lettering, and depicted Frech's pencil drawing of two gulls in flight over a sandy dune along the shore. Carson gave the first copy to her mother, who opened it and wept at the simple dedication, "To my mother." The second copy went to Elmer Higgins, inscribed, "To Mr. Higgins, who started it all." The jacket bore no photograph of the author, but included a sizable paragraph on her education and scientific experience.

The Scientific Book Club adopted *Under the Sea-Wind* as its December selection. Its reviewers established the tone of the critical notices, praising the depth of its information about sea life, noting "there is poetry here, but no false sentimentality." Other critics commented that it read like fiction but was in fact a scientifically accurate account of life in the ocean and along the shore. Carson was especially pleased by the opinions of other scientists and naturalists, including some of the leading fisheries biologists of the time, who usually had little patience with popularizations of science. William Beebe remarked on its lyrical beauty and faultless science in *The Saturday Review of Literature*. Even better, Beebe later included two chapters in his anthology of the best natural history writing, which began with Aristotle and ended with Carson. The best review, and the one that ultimately meant the most to Carson, however, did not appear until 1952. It was written by Henry Beston, the author of *The Outermost House*, a book Carson considered one of the greatest natural histories of the seashore of all time.

Poised for the popular reception that she had every reason to expect after such glowing critical evaluations, Carson's hopes were dashed by world events. Although she submitted the book to various prize-giving organizations, she was ultimately deprived of the commercial success she had counted on. Barely a month after publication, in December 1941 Japan attacked the American naval base at Pearl Harbor. Preparations for the United States' entry into World War II dominated the news and overshadowed everything else. Carson later recalled her disappointment at the outcome of her first commercial venture with wry humor. "The world received the event" she said, "with superb indifference. The rush to the book store that is the author's dream never materialized." Barely two thousand copies of *Under the Sea-Wind* had been sold when Simon & Schuster allowed it to go out of print in August 1946. Carson's royalties, including fees for use of extracts in other publications and a German edition, amounted to less than a thousand dollars. Carson turned her attention to her wartime government job at the U.S. Fish and Wildlife Service. Little wonder that she advised a friend to try writing magazine articles. "Except for the rare miracles where a book becomes a 'best seller,' I am convinced that writing a book is a very poor gamble financially."

A decade after the publication of *Under the Sea-Wind,* Carson again fulfilled her own criteria as the author of a new "best seller" on the sea. This time no world event eclipsed its publication, and *The Sea Around Us* (1951) broke publishing records, making its author the most authoritative voice of public science in America, and garnering international attention. Carson's financial success enabled her to buy back the publishing rights of her first book as well as the gravure negatives of Howard Frech's fine illustrations, which had been moldering in a warehouse. The critical success of *The Sea Around Us* prompted its publisher, Oxford University Press, to reissue *Under the Sea-Wind* on April 13, 1952, when Carson's public acclaim was at its height. Although the new edition was smaller in format and lacked Frech's drawings, the Book-of-the-Month Club chose it as an alternate selection in June 1952, and *Life* magazine published all of Part I, with striking illustrations by one of their staff artists. Nearly 40,000 copies of *Under the Sea-Wind* were sold even before its publication.

The New York Times noted a publishing achievement "as rare as a total solar eclipse." *Under the Sea-Wind* appeared on the best-seller list in tenth place, while *The Sea Around Us* vacillated between first and second. Not only did Carson feel vindicated by the popular reception of her once-neglected first book, she also took a certain delight in seeing how the book critics effused over what they had once ignored. "Once or twice in a generation does the world get a physical scientist with literary genius," the *Times* critic effused. "Miss Carson has written a classic in *The Sea Around Us. Under the Sea-Wind* may be another."

Modern readers will understand such extraordinary praise. The level of freshness that Carson brings to her account of the cycles of seasons and the struggle of each creature for survival marks *Sea-Wind* in some ways as her most successful book. Her voice is that of both scientist and poet, a writer in love with the wonder in nature that she has discovered and can share. The most moving passages of *Under the Sea-Wind* come from her own experience. "To stand at the edge of the sea, to feel the breath of a mist moving over a great salt marsh, to watch the flight of shore birds that have swept up and down the surf lines for untold millions of years, is to have knowledge of things that are as nearly eternal as any earthly life can be. These things were before ever man stood on the shore of the ocean and looked out upon it with wonder; they continue year in, year out, through the centuries and the ages, while kingdoms rise and fall." It is this reaffirmation of life that Rachel Carson's words still evoke—

the calming sense of continuity that comes from watching an eternal stream of teeming, struggling life move on—a stream of life in which death is only an incident. From Carson's response, we too can take hope in possibility.

Linda Lear
Bethesda, Maryland
February 2007

Suggestions for Further Reading

Brooks, Paul. *Rachel Carson at Work*. Los Angeles: Sierra Club Books, 1998.

Carson, Rachel. *The Edge of the Sea*. Boston: Houghton Mifflin, 1998.

______. *The Sea Around Us*. New York: Oxford University Press, 2003.

______. *The Sense of Wonder*. New York: HarperCollins, 1998.

Cramer, Deborah. *Great Waters*. New York: W. W. Norton & Company, 2001.

______. *Silent Spring*. Boston: Houghton Mifflin, 2007.

Lear, Linda (ed). *Lost Woods: The Discovered Writing of Rachel Carson*. Boston: Beacon Press, 1998.

______. *Rachel Carson: Witness for Nature*. New York: Henry Holt & Co., 1997.

Matthiessen, Peter (ed). *Courage for the Earth: Writers, Scientists, and Activists Celebrate the Life and Writing of Rachel Carson*. Boston: Houghton Mifflin, 2007.

Under the Sea-Wind

While the sun and the rain live, these shall be;
Till a last wind's breath upon all these blowing
Roll the sea.

—Swinburne

Foreword

Under the Sea-Wind was written to make the sea and its life as vivid a reality for those who may read the book as it has become for me during the past decade.

It was written, moreover, out of the deep conviction that the life of the sea is worth knowing. To stand at the edge of the sea, to sense the ebb and the flow of the tides, to feel the breath of a mist moving over a great salt marsh, to watch the flight of shore birds that have swept up and down the surf lines of the continents for untold thousands of years, to see the running of the old eels and the young shad to the sea, is to have knowledge of things that are as nearly eternal as any earthly life can be. These things were before ever man stood on the shore of the ocean and looked out upon it with wonder; they continue year in, year out, through the centuries and the ages, while man's kingdoms rise and fall.

In planning the book I was confronted at the very outset with the problem of a central character. It soon became evident that there was no single animal—bird, fish, mammal, or any of the sea's lesser creatures—that could live in all the various parts of the sea I proposed to describe. That problem was instantly solved, however, when I realized that the sea itself must be the central character whether I wished it or not; for the sense of the sea, holding the power of life and death over every one of its creatures from the smallest to the largest, would inevitably pervade every page.

Under the Sea-Wind is a series of descriptive narratives unfolding successively the life of the shore, the open sea, and the sea bottom. Because in the book the reader is an observer of events which are narrated with little or no comment, perhaps a few "program notes" are in order.

In Book One ("Edge of the Sea") I have re-created the life of a stretch of North Carolina sea coast—a place of rolling sand dunes where the sea oats grow, of wide, salty marshes, of quiet sounds, and wild ocean beach. I begin with the spring, when the black skimmers are returning from the south, shad are running in from the sea to the rivers, and the great spring migration of the shore birds is at its height. To see a sandpiper running and probing at the edge of the surf in the spring is to glimpse a migrant on the eve of an adventure so remarkable that I have given a whole chapter to the summer saga of the shore birds on the Arctic tundras. Then we return with the birds to the Carolina sound country at the end of the summer, and read in all the movements of birds, fish, shrimp, and other water creatures the record of the changing seasons.

Book Two ("The Gull's Way") is a parallel picture of the same period of time on the open ocean, but here the cycle of the seasons take a different form. The life of the open sea—miles beyond sight of land—is various, strangely beautiful, and wholly unknown to all but a fortunate few. Book Two is the story of a true sea rover—a mackerel—from birth in the great ocean nursery of the surface waters, through all the vicissitudes of early life among the drifting plankton herds and youth in a sheltered New England harbor, to membership in a wandering school of mackerel subject to the depredations of fish-eating birds, large fishes, and man.

For Book Three ("River and Sea") there remained the gently sloping sea bottom that forms the rim or shelf of the continent, the steep descent of the continental slopes, and finally the abyss or the deep sea. Fortunately there is one creature whose life embraces all these in a history without parallel in the annals of sea or land. This creature is the eel. To picture the whole life of this remarkable animal, however, it was necessary to begin in the far tributaries of a coastal river where the eels spend most of their adult life and to trace the spawning migration of the eels in autumn to the sea. Other fish run out of the bays and sounds in the fall, but only far enough to find warm water in which to winter.

But the eels go on, to a deep abyss near the Sargasso Sea, where they spawn and die. From this strange world of the deep sea the young eels return alone each spring to the coastal rivers.

To get the feeling of what it is like to be a creature of the sea requires the active exercise of the imagination and the temporary abandonment of many human concepts and human yardsticks. For example, time measured by the clock or the calendar means nothing if you are a shore bird or a fish, but the succession of light and darkness and the ebb and flow of the tides mean the difference between the time to eat and the time to fast, between the time an enemy can find you easily and the time you are relatively safe. We cannot get the full flavor of marine life—cannot project ourselves vicariously into it—unless we make these adjustments in our thinking.

On the other hand, we must not depart too far from analogy with human conduct if a fish, shrimp, comb jelly, or bird is to seem real to us—as real a living creature as he actually is. For these reasons I have deliberately used certain expressions which would be objected to in formal scientific writing. I have spoken of a fish "fearing" his enemies, for example, not because I suppose a fish experiences fear in the same way that we do, but because I think he *behaves as though he were frightened*. With the fish, the response is primarily physical; with us, primarily psychological. Yet if the behavior of the fish is to be understandable to us, we must describe it in the words that most properly belong to human psychological states.

In choosing names for the animals I have followed the plan of using, whenever possible, the scientific name for the genus to which each belongs. Where that name is too formidable I have substituted something descriptive of the appearance of the creature, or, in the case of some of the Arctic animals, have used the Eskimo names.

A glossary has been included at the end of the book to provide an introduction to little-known animals and plants of the sea, or to renew the reader's acquaintance with those already known to him.

No one person, even in a long lifetime, could become intimately acquainted through personal experience with every phase of the sea and its life. To supplement my own experience I have drawn freely upon a rich scientific and semipopular literature for basic facts upon which I have placed my own interpretation in weaving them into the story. To list all the sources on which I have drawn would be impossible, but a few outstanding examples are the following: Arthur Cleveland Bent's matchless series of thirteen volumes on the life histories of North American birds; Henry B. Bigelow's *Fishes of the Gulf of Maine, Plankton of the Gulf of Maine,* and various papers in scientific journals on his exploration of the coast waters from Maine to Cape Hatteras; Johannes Schmidt's monumental papers on the life history of the eel; George M. Sutton's *Exploration of Southampton Island;* unpublished manuscripts by O. E. Sette on the life history of the mackerel; and that bible of oceanography, *The Depths of the Ocean,* by Sir John Murray and Johan Hjort.

In addition to these printed sources, I have profited by contacts with various people who have placed portions of their rich experience with marine life at my disposal. First among these should be mentioned Elmer Higgins, but for whose interest, encouragement, and assistance the book might never have been written. Among others who have answered my questions patiently and helpfully are Robert A. Nesbit, William C. Neville, John C. Pearson, and Edward Bailey.

BOOK I

EDGE OF THE SEA

I
FLOOD TIDE

The Island lay in shadows only a little deeper than those that were swiftly stealing across the sound from the east. On its western shore the wet sand of the narrow beach caught the same reflection of palely gleaming sky that laid a bright path across the water from island beach to horizon. Both water and sand were the color of steel overlaid with the sheen of silver, so that it was hard to say where water ended and land began.

Although it was a small island, so small that a gull might have flown across it with a score of wing beats, night had already come to its northern and eastern end. Here the marsh grasses waded boldly out into dark water, and shadows lay thick among the low-growing cedars and yaupons.

With the dusk a strange bird came to the island from its nesting grounds on the outer banks. Its wings were pure black, and from tip to tip their spread was more than the length of a man's arm. It flew steadily and without haste across the sound, its progress as measured and as meaningful as that of the shadows which little by little were dulling the bright water path. The bird was called Rynchops, the black skimmer.

As he neared the shore of the island the skimmer drifted closer to the water, bringing his dark form into strong silhouette against the gray sheet, like the shadow of a great bird that passed unseen above. Yet so quietly did he approach that the sound of his wings, if sound there were, was lost in the whisper song of the water turning over the shells on the wet sand.

At the last spring tide, when the thin shell of the new moon brought the water lapping among the sea oats that fringed the dunes of the banks, Rynchops and his kin had arrived on the

outer barrier strip of sand between sound and sea. They had journeyed northward from the coast of Yucatán, where they had wintered. Under the warm June sun they would lay their eggs and hatch their buff-colored chicks on the sandy islands of the sound and on the outer beaches. But at first they were weary after the long flight and they rested by day on sand bars when the tide was out or roamed over the sound and its bordering marshes by night.

Before the moon had come to the full, Rynchops had remembered the island. It lay across a quiet sound from which the banks shouldered away the South Atlantic rollers. To the north the island was separated from the mainland by a deep gutter where the ebbing tides raced strongly. On the south side the beach sloped gently, so that at slack water the fishermen could wade out half a mile before the water came above their armpits as they raked scallops or hauled their long seines. In these shallows young fishes swarmed, feeding on the small game of the waters, and shrimp swam with backward flipping of their tails. The rich life of the shallows brought the skimmers nightly from their nesting grounds on the banks, to take their food from the water as they moved with winnowing flight above it.

About sunset the tide had been out. Now it was rising, covering the afternoon resting places of the skimmers, moving through the inlet, and flowing up into the marshes. Through most of the night the skimmers would feed, gliding on slender wings above the water in search of the small fishes that had moved in with the tide to the shelter of grassy shallows. Because they fed on the rising tide, the skimmers were called flood gulls.

On the south beach of the island, where water no deeper than a man's hand ran over the gently ribbed bottom, Rynchops began to wheel and quarter over the shallows. He flew with a curious, lilting motion, lifting his wings high after the downstroke. His head was bent sharply so that the long lower bill, shaped like a scissor blade, might cut the water.

The blade or cutwater plowed a miniature furrow over the placid sheet of the sound, setting up wavelets of its own and sending vibrations thudding down through the water to rebound from the sandy bottom. The wave messages were received by the blennies and killifish that were roving the shallows on the alert for food. In the fish world many things are told by sound waves. Sometimes the vibrations tell of food animals like small shrimps or oar-footed crustaceans moving in swarms overhead. And so at the passing of the skimmer the small fishes came nosing at the surface, curious and hungry. Rynchops, wheeling about, returned along the way he had come and snapped up three of the fishes by the rapid opening and closing of his short upper bill.

Ah-h-h-h, called the black skimmer. *Ha-a-a-a! Ha-a-a-a! Ha-a-a-a!* His voice was harsh and barking. It carried far across the water, and from the marshes there came back, like echoes, the answering cries of other skimmers.

While the water was reclaiming inch after inch of sandy shore, Rynchops moved back and forth over the south beach of the island, luring the fishes to rise along his path and seizing them on his return. After he had taken enough minnows to appease his hunger he wheeled up from the water with half a dozen flapping wing beats and circled the island. As he soared above the marshy eastern end, schools of killifish moved beneath him through the forests of sea hay, but they were safe from the skimmer, whose wingspread was too great to allow him to fly among the clumps of grass.

Rynchops swerved out around the dock that had been built by the fisherman who lived on the island, crossed the gutter, and swept far over the salt marshes, taking joy in flight and soaring motion. There he joined a flock of other skimmers and together they moved over the marshes in long lines and columns, sometimes appearing as dark shadows on the night sky; sometimes as spectral birds when, wheeling swallow-like in air, they showed white breasts and gleaming underparts. As they flew they raised their voices in the weird night chorus of the skimmers, a strange medley of notes high-pitched and low, now soft as the cooing of a mourning dove, and again harsh as the cawing of a crow; the whole chorus rising and falling, swelling and throbbing, dying away in the still air like the far-off baying of a pack of hounds.

The flood gulls circled the island and crossed and recrossed the flats to the southward. All through the hours of the rising tide, they would hunt in flocks over the quiet waters of the sound. The skimmers loved nights of darkness, and tonight thick clouds lay between the water and the moon's light.

On the beach the water was moving with soft tinkling sounds among the windrows of jingle shells and young scallop shells. It ran swiftly under heaps of sea lettuce to rouse sand fleas that had taken refuge there when the tide ebbed that afternoon. The beach hoppers floated out on the backlash of each wavelet and moved in the returning water, swimming on their backs, legs uppermost. In the water they were comparatively safe from their enemies the ghost crabs, who roamed the night beaches on swift and silent feet.

In the waters bordering the island many creatures besides the skimmers were abroad that night, foraging in the shallows. As the darkness grew and the incoming tide lapped higher and higher among the marsh grasses, two diamondback terrapins slipped into the water to join the moving forms of others of their kind. These were females, who had just finished laying their eggs above the high-tide line. They had dug nests in the soft sand, working with hind feet until they scooped out jug-shaped holes not quite so deep as their own bodies were long. Then they had deposited their eggs, one five, the other eight. These they had carefully covered with sand, crawling back and forth to conceal the location of the nest. There were other nests in the sand, but none more than two weeks old, for May is the beginning of the nesting season among the diamondbacks.

As Rynchops followed the killifish in toward the shelter of the marsh he saw the terrapins swimming in the shallow water where the tide was moving swiftly. The terrapins nibbled at the marsh grasses and picked off small coiled snails that had crept up the flat blades. Sometimes they swam down to take crabs off the bottom. One of the two terrapins passed between two slender uprights like stakes thrust into the sand. They were the legs of the solitary great blue heron who flew every night from his rookery three miles away to fish from the island.

The heron stood motionless, his neck curved back on his shoulders, his bill poised to spear fish as they darted past his legs. As the terrapin moved out into deeper water she startled a young mullet and sent it racing toward the beach in confusion and panic. The sharp-eyed heron saw the movement and with a quick dart seized the fish crosswise in his bill. He tossed it into the air, caught it headfirst, and swallowed it. It was the first fish other than small fry that he had caught that night.

The tide was almost halfway to the confused litter of sea wrack, bits of sticks, dried claws of crabs, and broken shell fragments that marked high-water level. Above the tide line there were faint stirrings in the sand where the terrapins had lately begun to lay their eggs. The season's young would not hatch until August, but many young of the year before still were buried in the sand, not yet roused from the torpor of hibernation. During the winter the young terrapins had lived on the remnant of yolk left from embryonic life. Many had died, for the winter had been long and the frosts had bitten deep into the sands. Those that survived were weak and emaciated, their bodies so shrunken within the shells that they were smaller than when they had hatched. Now they were moving feebly in the sands where the old terrapins were laying the eggs of a new generation of young.

About the time the tide was midway to the flood, a wave of motion stroked the tops of the grasses above the terrapin egg bed, as though a breeze passed, but there was little wind that night. The grasses above the sand bed parted. A rat, crafty with the cunning of years and filled with the lust for blood, had come down to the water along a path which his feet and his thick tail had worn to a smooth track through the grass. The rat lived with his mate and others of his kind under an old shed where the fisherman kept his nets, faring well on the eggs of the many birds that nested on the island, and on the young birds.

As the rat looked out from the fringe of grass bordering the terrapin nests the heron sprang from the water a stone's throw away with a strong flapping of his wings and flew across the island to the north shore. He had seen two fishermen in a small boat coming around the western tip of the island. The fishermen had been gigging flounders, spearing them on the bottom in shallow water by the light of a torch which flared at the bow. A yellow splotch of light moved over the dark water in advance of the boat and sent trembling streamers across the wavelets that rippled shoreward from the boat's passing. Twin points of green fire glowed in the grass above the sand bed. They remained stationary until the boat had passed on around the south shore and had headed toward the town docks. Only then did the rat glide down from the path onto the sand.

The scent of terrapin and of terrapin eggs, fresh laid, was heavy in the air. Snuffling and squeaking in excitement, the rat began to dig and in a few minutes had uncovered an egg, had pierced the shell, and had sucked out the yolk. He then uncovered two other eggs and might have eaten them if he had not heard a movement in a nearby clump of marsh grass—the scrambling of a young terrapin struggling to escape the water that was seeping up around its tussock of tangled roots and mud. A dark form moved across the sand and through the rivulet of water. The rat seized the baby terrapin and carried it in his teeth through the marsh grasses to a hummock of higher ground. Engrossed in gnawing away the thin shell of the terrapin, he did not notice how the tide was creeping up about him and running deeper around the hummock. It was thus that the blue heron, wading back around the shore of the island, came upon the rat and speared him.

There were few sounds that night except those of the water and the water birds. The wind was asleep. From the direction of the inlet there came the sound of breakers on the barrier beach, but the distant voice of the sea was hushed almost to a sigh, a sort of rhythmic exhalation as though the sea, too, were asleep outside the gates of the sound.

It would have taken the sharpest of ears to catch the sound of a hermit crab dragging his shell house along the beach just above the water line: the elfin shuffle of his feet on the sand, the sharp grit as he dragged his own shell across another; or to have discerned the spattering tinkle of the tiny droplets that fell when a shrimp, being pursued by a school of fish, leaped clear of the water. But these were the unheard voices of the island night, of the water and the water's edge.

The sounds of the land were few. There was a thin insect tremolo, the spring prelude to the incessant chiton fiddles that later in the season would salute the night. There was the murmur of sleeping birds in the cedars—jackdaws and mockingbirds—who now and again roused enough to twitter drowsily one to another. About midnight a mockingbird sang for almost a quarter of an hour, imitating all the bird songs he had heard that day and adding trills, chuckles, and whistles all his own. Then he, too, subsided and left the night again to the water and its sounds.

There were many fish moving in through the deep water of the channel that night. They were full-bellied fish, soft-finned and covered with large silvery scales. It was a run of spawning shad, fresh from the sea. For days the shad had lain outside the line of breakers beyond the inlet. Tonight with the rising tide they had moved in past the clanging buoy that guided fishermen returning from the outer grounds, had passed through the inlet, and were crossing the sound by way of the channel.

As the night grew darker and the tides pressed farther into the marshes and moved higher into the estuary of the river, the silvery fish quickened their movements, feeling their way along the streams of less saline water that served them as paths to the river. The estuary was broad and sluggish, little more than an arm of the sound. Its shores were ragged with salt marsh, and far up along the winding course of the river the pulsating tides and the bitter tang of the water spoke of the sea.

Some of the migrating shad were three years old and were returning to spawn for the first time. A few were a year older and were making their second trip to the spawning grounds up the river. These were wise in the ways of the river and of the strange crisscross shadows it sometimes contained.

By the younger shad the river was only dimly remembered, if by the word "memory" we may call the heightened response of the senses as the delicate gills and the sensitive lateral lines perceived the lessening saltiness of the water and the changing rhythms and vibrations of the inshore waters. Three years before they had left the river, dropping downstream to the estuary as young fish scarcely as long as a man's finger, moving out to sea with the coming of autumn's chill. The river forgotten, they roamed widely in the sea, feeding on shrimps and amphipods. So far and so deviously did they travel that no man could trace their movements. Perhaps they wintered in deep, warm water far below the surface, resting in the dim twilight of the continent's edge, making an occasional timid journey out over the rim beyond which lay only the blackness and stillness of the deep sea. Perhaps in summer they roved the open ocean, feeding on the rich life of the surface, packing layers of white muscle and sweet fat beneath their shining armor of scales.

The shad roamed the sea paths known and followed only by fish while the earth moved three times through the cycle of the zodiac. In the third year, as the waters of the sea warmed slowly to the southward-moving sun, the shad yielded to the promptings of race instinct and returned to their birthplaces to spawn.

Most of the fish coming in now were females, heavy with unshed roe. It was late in the season and the largest runs had gone before. The bucks, who came into the river first, were already on the spawning grounds, as were many of the roe shad. Some of the early-run fish had pressed upstream as far as a hundred miles to where the river had its formless beginnings in dark cypress swamps.

Each of the roe fish would shed in a season more than a hundred thousand eggs. From these perhaps only one or two young would survive the perils of river and sea and return in time to spawn, for by such ruthless selection the species are kept in check.

The fisherman who lived on the island had gone out about nightfall to set the gill nets that he owned with another fisherman from the town. They had anchored a large net almost at right angles to the west shore of the river and extending well out into the stream. All the local fishermen knew from their fathers, who had it from their fathers, that shad coming in from the channel of the sound usually struck in toward the west bank of the river when they entered the shallow estuary, where no channel was kept open. For this reason the west bank was crowded with fixed fishing gear, like pound nets, and the fishermen who operated movable gear competed bitterly for the few remaining places to set their nets.

Just above the place where the gill net had been set tonight was the long leader of a pound net fixed to posts driven into the soft bottom. The year before there had been a fight when the fishermen who owned the pound nets had discovered the gill netters taking a good catch of shad from their own net, which they had set directly downstream from the pound, heading off most of the fish. The gill-net fishermen were outnumbered, and for the rest of the season had fished in another part of the estuary, making poor catches and cursing the pound netters. This year they had tried setting the nets at dusk and returning to fish them by daybreak. The rival fishermen did not tend the pound till about sunrise, and by that time the gill netters were always downstream again, nets in their boat, nothing to prove where they had been fishing.

About midnight, as the tide neared the full, the cork line bobbed as the first of the migrating shad struck the gill net. The line vibrated and several of the cork floats disappeared under the water. The shad, a four-pound roe, had thrust her head through one of the meshes of the net and was struggling to free herself. The taut circle of twine that had slipped under the gill covers cut deeper into the delicate gill filaments as the fish lunged against the net; lunged again to free herself from something that was like a burning, choking collar; something that held her in an invisible vise and would neither let her go on upstream nor turn and seek sanctuary in the sea she had left.

The cork line bobbed many times that night and many fish were gilled. Most of them died slowly of suffocation, for the twine interfered with the rhythmic respiratory movements of the gill covers by which fish draw streams of water in through the mouth and pass them over the gills. Once the line bobbed very hard and for ten minutes was pulled below the surface. That was when a grebe, swimming fast five feet under water after a fish, went through the net to its shoulders and in its violent struggles with wings and lobed feet became hopelessly entangled. The grebe soon drowned. Its body hung limply from the net, along with a score of silvery fish bodies with heads pointing upstream in the direction of the spawning grounds where the early-run shad awaited their coming.

By the time the first half-dozen shad had been caught in the net, the eels that lived in the estuary had become aware that a feast was in the offing. Since dusk they had glided with sinuating motion along the banks, thrusting their snouts into crabholes and seizing whatever they could catch in the way of small water creatures. The eels lived partly by their own industry but were also robbers who plundered the fishermen's gill nets when they could.

Almost without exception the eels of the estuary were males. When the young eels come in from the sea, where they are born, the females press far up into rivers and streams, but the males wait about the river mouths until their mates-to-be, grown sleek and fat, rejoin them for the return journey to the sea.

As the eels poked their heads out of the holes under the roots of the marsh grasses and swayed gently back and forth, savoring eagerly the water that they drew into their mouths, their keen senses caught the taste of fish blood, which was diffusing slowly through the water as the gilled shad struggled to escape. One by one they slipped out of their holes and followed the taste trail through the water to the net.

The eels feasted royally that night, since most of the fish caught by the net were roe shad. The eels bit into the abdomens with sharp teeth and ate out the roe. Sometimes they ate out all the flesh as well, so that nothing remained but a bag of skin, with an eel or two inside. The marauders could not catch a live shad free in the river, so their only chance for such a meal was to rob the gill nets.

As the night wore on and the tide passed the flood, fewer shad came upstream and no more were caught by the gill net. A few of those that had been caught and insecurely gilled just before the tide turned were released by the return flow of water to the sea. Of those that escaped the gill net, some had been diverted by the leader of the pound net and had followed along the walls of small-meshed netting into the heart of the pound net and thence into the pocket, where they were trapped; but most had gone on upstream for several miles and were resting now until the next tide.

The posts of the wharf on the north shore of the island showed two inches of wet watermark when the fisherman came down with a lantern and a pair of oars. The silence of the waiting night was broken by the thud of his boots on the wharf; the grating of oars fitting into oarlocks; the splash of water from the oars as he pulled out into the gutter and headed toward the town docks to pick up his partner. Then the island settled to silence again and to waiting.

Although there was as yet no light in the east, the blackness of water and air was perceptibly lessening, as though the darkness that remained were something less solid and impenetrable than that of midnight. A freshening air moved across the sound from the east and, blowing across the receding water, sent little wavelets splashing on the beach.

Most of the black skimmers had already left the sound and returned by way of the inlet to the outer banks. Only Rynchops remained. Seemingly he would never tire of circling the island, of making wide sorties out over the marshes or up the estuary of the river where the shad nets were set. As he crossed the gutter and started up the estuary once more, there was enough light to see the two fishermen maneuvering their boat into position beside the cork line of the gill net. White mist was moving over the water and swirling around the fishermen, who were standing in their boat and straining to raise the anchor line at the end of the net. The anchor came up, dragging with it a clump of widgeon grass, and was dropped in the bottom of the boat.

The skimmer passed upstream about a mile, flying low to the water, then turned by circling widely over the marshes and came down to the estuary again. There was a strong smell of fish and of water weeds in the air, which came to him through the morning mists, and the voices of the fishermen were borne clearly over the water. The men were cursing as they worked to raise the gill net, disentangling the fish before they piled the dripping net on the flat bottom of the skiff.

As Rynchops passed about half a dozen wing beats from the boat, one of the fishermen flung something violently over his shoulder—a fish head with what looked like a stout white cord attached. It was the skeleton of a fine roe shad, all that remained, save the head, after the feast of the eels.

The next time Rynchops flew up the estuary he met the fishermen coming downstream on the ebbing tide, net piled in the boat over some half-dozen shad. All the others had been gutted or reduced to skeletons by the eels. Already gulls were gathering on the water where the gill net had been set, screaming their pleasure over the refuse which the fishermen had thrown overboard.

The tide was ebbing fast, surging through the gutter and running out to sea. As the sun's rays broke through the clouds in the east and sped across the sound, Rynchops turned to follow the racing water seaward.

2
SPRING FLIGHT

The night when the great run of shad was passing through the inlet and into the river estuary was a night, too, of vast movements of birds into the sound country.

At daybreak and the half tide two small sanderlings ran beside the dark water on the ocean beach of the barrier island, keeping in the thin film at the edge of the ebbing surf. They were trim little birds in rust and gray plumage, and they ran with a twinkle of black feet over the hard-packed sand, where puffs of blown spume or sea froth rolled like thistledown. They belonged to a flock of several hundred shore birds that had arrived from the south during the night. The migrants had rested in the lee of the great dunes while darkness remained; now growing light and ebbing water were drawing them to the sea's edge.

As the two sanderlings probed the wet sand for small, thin-shelled crustaceans, they forgot the long flight of the night before in the excitement of the hunt. For the moment they forgot, too, that faraway place which they must reach before many days had passed—a place of vast tundras, of snow-fed lakes, and of midnight sun. Blackfoot, leader of the migrant flock, was making his fourth journey from the southernmost tip of South America to the Arctic nesting grounds of his kind. In his short lifetime he had traveled more than sixty thousand miles, following the sun north and south across the globe, some eight thousand miles spring and fall. The little hen sanderling that ran beside him on the beach was a yearling, returning for the first time to the Arctic she had left as a fledgling nine months before. Like the older sanderlings, Silverbar had changed her winter plumage of pearly gray for a mantle heavily splashed with cinnamon and rust, the colors worn by all sanderlings on their return to their first home.

In the fringe of the surf, Blackfoot and Silverbar sought the sand bugs or Hippa crabs that honeycombed the ocean beach with their burrowings. Of all the food of the tide zone they loved best these small, egg-shaped crabs. After the retreat of each wave the wet sand bubbled with the air released from the shallow crab burrows, and a sanderling could, if he were quick and sure of foot, insert his bill and draw out the crab before the next wave came tumbling in. Many of the crabs were washed out by the swift rushes of the waves and left kicking in liquefying sand. Often the sanderlings seized these crabs in the moment of their confusion, before they could bury themselves by furious scrambling.

Pressing close to the backwash, Silverbar saw two shining air bubbles pushing away the sand grains and she knew that a crab was beneath. Even as she watched the bubbles her bright eyes saw that a wave was taking form in the tumbling confusion of the surf. She gauged the speed of the mound of water as it ran, toppling, up the beach. Above the deeper undertones of moving water she heard the lighter hiss that came as the crest began to spill. Almost in the same instant the feathered antennae of the crab appeared above the sand. Running under the very crest of the green water hill, Silverbar probed vigorously in the wet sand with opened bill and drew out the crab. Before the water could so much as wet her legs she turned and fled up the beach.

While the sun still came in level rays across the water, others of the sanderling flock joined Blackfoot and Silverbar and the beach was soon dotted with small shore birds.

A tern came flying along the surf line, his black-capped head bent and his eyes alert for the movement of fish in the water. He watched the sanderlings closely, for sometimes a small beach bird could be frightened into giving up its catch. When the tern saw Blackfoot run swiftly into the path of a wave and seize a crab he slanted down menacingly, screaming threats in a shrill, grating voice.

Tee-ar-r-r! Tee-ar-r-r! rattled the tern.

The swoop of the white-winged bird, which was twice as large as the sanderling, took Blackfoot by surprise, for his senses had been occupied with eluding the onrush of water and preventing the escape of the large crab held in his bill. He sprang into the air with a sharp *Keet! Keet!* and circled out over the surf. The tern whirled after him in pursuit, crying loudly.

In his ability to bank and pivot in the air Blackfoot was fully the equal of the tern. The two birds, darting and twisting and turning, coming up sharply together and falling away again into the wave troughs, passed out beyond the breakers and the sound of their voices was lost to the sanderling flock on the beach.

As he rose steeply into the air in pursuit of Blackfoot, the tern caught sight of a glint of silver in the water below. He bent his head to mark the new prey more certainly and saw the green water spangled with silver streaks as the sun struck the flanks of a school of feeding silversides. Instantly the tern tipped his body steeply into a plane perpendicular to the water. He fell like a stone, although his body could not have weighed more than a few ounces, struck the water with a splash and a shower of spray, and in a matter of seconds emerged with a fish curling in his bill. By this time Blackfoot, forgotten by the tern in the excitement engendered by the bright flashes in the water, had reached the shore and dropped down among the feeding sanderlings, where he was running and probing busily as before.

After the tide turned, the water pressed stronger from the sea. The waves came in with a deeper swell and a heavier crash, warning the sanderlings that feeding on the ocean beach was no longer safe. The flock wheeled out over the sea, with a flashing of the white wing bars that distinguished them from other sandpipers. They flew low over the crests of the waves as they traveled up the beach. So they came to the point of land called Ship's Shoal, where the sea had broken through the barrier island to the sound years before.

At the point the inlet beach lay level as a floor from the sea on the south side to the sound on the north. The wide sand flat was a favorite resting place for sandpipers, plovers, and other shore birds; and it was loved, too, by the terns, the skimmers, and the gulls, who make their living from the sea, but gather to rest on shores and sandspits.

That morning the inlet beach was thronged with birds, resting and waiting for the tide turn that they might feed again, fueling small bodies for the northward journey. It was the month of May, and the great spring migration of the shore birds was at its height. Weeks before, the waterfowl had left the sounds. Two spring tides and two neap tides had passed since the last skein of snow geese had drifted to the north, like wisps of cloud in the sky. The mergansers had gone in February, looking for the first breaking up of the ice in the northern lakes, and soon after them the canvasbacks had left the wild celery beds of the estuary and followed the retreating winter to the north. So, too, the brant, eaters of the eel grass that carpeted the shallows of the sound, the swift blue-winged teal, and the whistling swans, filling the skies with their soft trumpetings.

Then the bell notes of the plovers had begun to ring among the sand hills and the liquid whistle of the curlew throbbed in the salt marshes. Shadowy forms moved through the night skies and pipings so soft as barely to be audible drifted down to fishing villages sleeping below, as the birds of shore and marsh poured northward along ancestral air lanes, seeking their nesting places.

Now while the shore birds slept on the inlet beach the sands belonged to other hunters. After the last bird had settled to rest, a ghost crab came out of his burrow in the loose white sand above the high-water mark. He sped along the beach, running swiftly on the tips of his eight legs. He paused at a mass of sea wrack left by the night tide not a dozen paces from the spot where Silverbar stood on the edge of the sanderling flock. The crab was a creamy tan, matching the sand so closely that he was all but invisible when he stood still. Only his eyes, like two black shoe buttons on stalks, showed color. Silverbar saw the crab crouch behind the litter of sea-oats stubble, leaves of beach grass, and pieces of sea lettuce. He was waiting for a sand hopper or beach flea to show itself by an unwary movement. As the ghost crabs knew, the beach fleas hid in the seaweeds when the tide was out, browsing on them and picking up bits of decaying refuse.

Before the tide had risen another hand's breadth, a beach flea crept out from under a green frond of sea lettuce and leaped with an agile flexing of its legs across a stem of sea oats, as large to it as a fallen pine. The ghost crab sprang like a pouncing cat and seized the flea in his large crushing claw, or chela, and devoured it. During the next hour he caught and ate many of the beach hoppers, stealing on silent feet from one vantage point to another as he stalked his prey.

After an hour the wind changed and blew in across the inlet channel, obliquely from the sea. One by one the birds shifted their position so that they faced the wind. Above the surf at the point they saw a flock of several hundred terns fishing. A shoal of small silvery fish was passing seaward around the point, and the air was filled with the white wing flashes of diving terns.

At intervals the birds on the beach at Ship's Shoal heard the flight music of hurrying flocks of black-bellied plovers, high in the sky; and twice they saw long lines of dowitchers passing northward.

At noon white wings sailed over the sand dunes and a snowy egret swung down onto long black legs. The bird alighted at the margin of a pond that lay, half encircled by marsh, between the eastern end of the dunes and the inlet beach. The pond was called Mullet Pond, a name given to it years before when it had been larger and mullet had sometimes come into it from the sea. Every day the small white heron came to fish the pond, seeking the killifish and other minnows that darted in its shallows. Sometimes, too, he found the young of larger fishes, for the highest tides of each month cut through the beach on the ocean side and brought in fish from the sea.

The pond slept in noonday quiet. Against the green of the marsh grass the heron was a snow-white figure on slim black stilts, tense and motionless. Not a ripple nor the shadow of a ripple passed beneath his sharp eyes. Then eight pale minnows swam single file above the muddy bottom, and eight black shadows moved beneath them.

With a snakelike contortion of its neck, the heron jabbed violently, but missed the leader of the solemn little parade of fish. The minnows scattered in sudden panic as the clear water was churned to muddy chaos by the feet of the heron, who darted one way and another, skipping and flapping his wings in excitement. In spite of his efforts, he captured only one of the minnows.

The heron had been fishing for an hour and the sanderlings, sandpipers, and plovers had been sleeping for three hours when a boat's bottom grated on the sound beach near the point. Two men jumped out into the water and made ready to drag a haul seine through the shallows on the rising tide. The heron lifted his head and listened. Through the fringe of sea oats on the sound side of the pond he saw a man walking down the beach toward the inlet. Alarmed, he thrust his feet hard against the mud and with a flapping of wings took off over the dunes toward the heron rookery in the cedar thickets a mile away. Some of the shore birds ran twittering across the beach toward the sea. Already the terns were milling about overhead in a noisy cloud, like hundreds of scraps of paper flung to the wind. The sanderlings took flight and crossed the point, wheeling and turning almost as one bird, and passed down the ocean beach about a mile.

The ghost crab, still at his hunting of beach fleas, was alarmed by the turmoil of birds overhead, by the many racing shadows that sped over the sand. By now he was far from his own burrow. When he saw the fisherman walking across the beach he dashed into the surf, preferring this refuge to flight. But a large channel bass was lurking nearby, and in a twinkling the crab was seized and eaten. Later in the same day, the bass was attacked by sharks and what was left of it was cast up by the tide onto the sand. There the beach fleas, scavengers of the shore, swarmed over it and devoured it.

Twilight found the sanderlings resting again at the point of land called Ship's Shoal, listening to the soft roar of wings in the air about them as the curlews came in from the salt marshes to roost for the night on the inlet beach. Silverbar crouched close to some of the older sanderlings because of the strange sounds and the movements of so many large birds. There must have been thousands of curlews. For an hour after dark they were arriving, in long V formations and dense flocks. Every year the big brown birds with the sickle-shaped bills stopped on their northward migration to feed on the fiddler crabs of the mud flats and marshes.

A stone's throw away several crabs, no larger than a man's thumbnail, moved across the beach, but the sound of their feet was like the sound of sand grains disturbed by the wind, and so not even Silverbar, who was resting near the edge of the sanderling flock, heard them passing. They waded into the shallows and let the cool water bathe their bodies. This had been a day of distress and terror for the fiddlers, with all the marshes filled with curlews. Many times each hour the shadow of a bird soaring down to alight in the marsh or the sight of one of the curlews walking down along the water's edge had sent the small crabs scattering like a herd of stampeding cattle. Then the hundreds of feet on the sand had made a sound like the rattling of stiff sheets of paper. As many as could had darted into burrows—their own burrows—any burrows they could reach. But the long, oblique tunnels in the sand had been poor sanctuary, for the curving bills of the curlews could probe them deeply.

Now with the grateful twilight the fiddler herds had moved down to the water line to search for food among the sand windrows left by the receding tide. With their little spoon claws the crabs felt busily among the sand grains, sorting out the microscopic cells of algae.

The crabs that had waded into the water were females carrying eggs on the broad aprons of their abdomens. Because of the egg masses they moved awkwardly and were unable to run from their enemies, and so all day they had remained hidden deep in the burrows. Now they swayed to and fro in the water, seeking to rid themselves of their burdens. This was an instinct that served to aerate the eggs adhering to the mother's body like bunches of miniature purple grapes. Although the season was early, some of the fiddlers carried gray egg masses, signifying that the young were ready for life. For these crabs the evening ritual of washing brought on the hatching of the eggs. With each movement of the mothers' bodies, many eggshells burst and clouds of larvae were hurled into the water. Even the killifish that were nibbling algae from the shells in the quiet shallows of the sound scarcely noticed the throngs of newborn creatures that drifted by, for any of the baby crabs thus abruptly released from the confining sphere of the egg could have passed through the eye of a needle.

The clouds of larvae were carried away on the still-ebbing tide and swept out through the inlet. When the first light should steal across the water they would find themselves in the strange world of the open sea, amid many perils which they must surmount, alone and unaided save for the self-protective instincts with which each was endowed at birth. Many would fail. The others, after long weeks of adventurous living, would put in to some distant shore, where the tides spread abundant feasts for fiddler crabs and marsh grasses offered home and shelter.

The night was noisy with the barking cries of the black skimmers who chased each other in play over the inlet, where the moon struck a white path across the water. The sanderlings had often seen the skimmers in South America, for many of them wintered as far south as Venezuela and Colombia. The skimmers, compared with the sanderlings, were birds of the tropics and knew nothing of the white world to which the shore birds were bound.

At intervals throughout the night the calls of Hudsonian curlews, migrating at a great height, came down from the sky. The curlews sleeping on the beach stirred uneasily and sometimes answered the cries with plaintive whistles.

It was the night of the full moon, the moon of the spring tides when the water presses far into the marshes and laps at the floorboards of fishermen's wharfs, and makes boats strain at their anchors.

The sea, that gleamed with the moon's lambent silver, drew to its surface many squids, dazed and fascinated by the light. The squids drifted on the sea, their eyes fixed on the moon. Gently they drew in water and expelled it in jets, propelling themselves backward away from the light at which they gazed. Moon-bewildered, their senses did not warn them that they were drifting into dangerous shoals until the harsh grate of sand brought sharp awakening. As they stranded, the hapless squids pumped water all the harder, driving themselves out of even the thinnest film, onto sand from which all water had ebbed away.

In the morning the sanderlings, moving down to the surf line to feed in the first light, found the inlet beach littered with dead squids. The sanderlings did not linger on this part of the beach, for although it was very early in the morning many large birds had gathered and were quarreling over the squids. They were herring gulls, bound from the Gulf Coast to Nova Scotia. They had been long delayed by stormy weather and they were ravenous. A dozen black-headed laughing gulls came and hovered, mewing, over the beach, dangling their feet as though to alight, but the herring gulls drove them away with fierce screaming and jabs of their bills.

By midday, with the rising tide, a strong wind was blowing in from the sea and storm clouds ran before it. The green ranks of the marsh grasses swayed and their tips bent to touch the rising water. After the first quarter of the tide rise all the marshes stood deep in water. The scattered sand shoals of the sound, favorite resting places of the gulls, were covered as the spring tide ran with the wind's weight behind it.

The sanderlings, along with flocks of other shore birds, took refuge close beneath the landward slopes of the dunes. There the forests of beach grass sheltered them. From their haven they saw the flock of herring gulls sweeping like a gray cloud over the vivid green of the marshes. The flock constantly changed shape and direction as it rolled, the leaders hesitating over a possible resting place, the laggards surging forward. Now they settled on a sand shoal, shrunk to a tenth the size it had been that morning. The water was rising. On they moved, to hover, fluttering and screaming, above a reef of oyster shell, where the water streamed neck-deep to a gull. At last the whole flock veered around and fought its way back into the face of the gale, coming to rest near the sanderlings in the shelter of the dunes.

Stormbound, all the migrants waited, unable to feed because of the heavy surf. At sea, out beyond the sheltering capes, a violent storm was raging. On the ocean beach two small birds, dazed and sick with buffeting, staggered over the sand, fell, and staggered on again. Land was to them a strange realm. Except for a short period each year when they visited small islands in the Antarctic Sea to rear their young, their world was of sky and rolling water. They were Wilson's petrels or Mother Carey's chickens, blown in by the storm from miles at sea. And once during the afternoon a dark brown bird with slender wings and hawklike bill came beating its way over the dunes and across the sound. Blackfoot the sanderling and many of the other shore birds crouched in terror, recognizing an ancestral enemy, the scourge of the northern breeding grounds. Like the petrels, the jaeger had ridden in on the gale from the open sea.

Before sunset, the skies lightened and the wind abated. While it was yet light the sanderlings left the barrier island and set out across the sound. Beneath them as they wheeled over the inlet was the deep green ribbon of the channel that wound, with many curvings, across the lighter shallows of the sound. They followed the channel, passing between the leaning red spar buoys, past the tide rips where the water streamed, broken into swirls and eddies, over a sunken reef of oyster shell, and came at last to the island. There they joined a company of several hundred white-rumped sandpipers, least sandpipers, and ring-necked plovers that were resting on the sand.

While the tide was still ebbing, the sanderlings fed on the island beach, but settled to rest before the arrival, at dusk, of Rynchops the black skimmer. As they slept, and as the earth rolled from darkness toward light, birds from many feeding places along the coast were hurrying along the flyways that lead to the north. For with the passing of the storm the air currents came fresh again and the wind blew clean and steady from the southwest. All through the night the cries of curlews and plovers and knots, of sandpipers and turnstones and yellowlegs, drifted down from the sky. The mockingbirds who lived on the island listened to the cries. The next day they would have many new notes in their rippling, chuckling songs to charm their mates and delight themselves.

About an hour before dawn the sanderling flock gathered together on the island beach, where the gentle tide was shifting the windrows of shells. The little band of brown-mottled birds mounted into the darkness and, as the island grew small beneath them, set out toward the north.

3
ARCTIC RENDEZVOUS

Winter still gripped the northland when the sanderlings arrived on the shores of a bay shaped like a leaping porpoise, on the edge of the frozen tundras of the barren grounds. They were among the first to arrive of all the migrant shore birds. Snow lay on the hills and drifted deep in the stream valleys. The ice was yet unbroken in the bay, and on the ocean shore it was piled in green and jagged heaps that moved, straining and groaning, with the tides.

But the lengthening days filled with sun had already begun to melt the snow on the south slopes of the hills, and on the ridges the wind had helped wear the snow blanket thin. There the brown of earth and the silver gray of reindeer moss showed through, and now for the first time that season the sharp-hoofed caribou could feed without pawing away the snow. At noon the white owls beating across the tundra beheld their own reflections in many small pools among the rocks, but by midafternoon the water mirrors were clouded with frost.

Already the rusty feathers were showing about the necks of the willow ptarmigans and brown hairs had appeared on the white coats of the foxes and weasels. Snow buntings hopped about in flocks that grew day by day, and the buds on the willows swelled and showed the first awakening of color under the sunshine.

There was little food for the migrant birds—lovers of warm sun and green, tossing surf. The sanderlings gathered miserably under a few dwarf willows that were sheltered from the northwest winds by a glacial moraine. There they lived on the first green buds of the saxifrage and awaited the coming of thaws to release the rich animal food of the Arctic spring.

But winter was yet to die. The second sun after the sanderlings' return to the Arctic burned dimly in the murky air. The clouds thickened and rolled between the tundra and the sun, and by midday the sky was heavy with unfallen snow. Wind came in over the open sea and over the ice packs, carrying a bitter air that turned to mist as it moved, swirling, over the warmer plains.

Uhvinguk, the lemming mouse who yesterday had sunned himself with many of his fellows on the bare rocks, ran into the burrows, winding tunnels in the deep, hard drifts, and to the grass-lined chambers where the lemmings dwelt in warmth even in midwinter. In the twilight of that day a white fox paused above the lemming burrow and stood with lifted paw. In the silence his sharp ears caught the sound of small feet along the runways below. Many times that spring the fox had dug down through the snow into these burrows and seized as many lemmings as he could eat. Now he whined sharply and pawed a little at the snow. He was not hungry, having killed and eaten a ptarmigan an hour before when he had come upon it, in a willow thicket, snipping off twigs; so today he only listened, perhaps to reassure himself that the weasels had not raided the lemming colony since his last visit. Then he turned and ran on silent feet along the path made by many foxes, not even pausing to glance at the sanderlings huddled in the lee of the moraine, and passed over the hill to the distant ridge where a colony of thirty small white foxes had their burrows.

Late that night, about the time the sun must have been setting somewhere behind the thick cloud banks, the first snow fell. Soon the wind rose and poured across the tundra like a flood of icy water that penetrated the thickest feathers and the warmest fur. As the wind came down shrieking from the sea, the mists fled before it across the barrens, but the snow clouds were thicker and whiter than the mists had been.

Silverbar, the young hen sanderling, had not seen snow since she had left the Arctic nearly ten months before to follow the sun southward toward the limit of its orbit, to the grasslands of the Argentine and the shores of Patagonia. Almost her whole existence had been of sun and wide white beaches and rippling green pampas. Now, crouched under the dwarf willows, she

could not see Blackfoot through the swirling whiteness, although she could have reached his side with a quick run of twenty paces. The sanderlings faced into the blizzard, as shore birds everywhere face into the wind. They huddled close together, wing to wing, and the warmth of their bodies kept the tender feet from freezing as they crouched on them.

If the snow had not drifted so, that night and all the next day, the loss of life would have been less. But the stream valleys filled up, inch by inch, throughout the night, and against the ridges the white softness piled deeper. Little by little, from the ice-strewn sea edge across miles of tundra, even far south to the fringe of the forests, the undulating hills and the ice-scoured valleys were flattening out, and a strange world, terrifying in its level whiteness, was building up. In the purple twilight of the second day the fall slackened, and the night was loud with the crying of the wind, but with no other voice, for no wild thing dared show itself.

The snow death had taken many lives. It had visited the nest of two snowy owls in a ravine that cut a deep scar in the hillside, near the willow copse that sheltered the sanderlings. The hen had been brooding the six eggs for more than a week. During the first night of wild storm the snow had drifted deep about her, leaving a round depression like a stream-bed pothole in which she sat. All through the night the owl remained on the nest, warming the eggs with her great body, that was almost furry in its plumage. By morning the snow was filling in around the feather-shod talons and creeping up around her sides. The cold was numbing, even through the feathers. At noon, with flakes like cotton shreds still flying in the sky, only the owl's head and shoulders were free of the snow. Several times that day a great form, white and silent as the snowflakes, had drifted over the ridge and hovered above the place where the nest was. Now Ookpik, the cock owl, called to his mate with low, throaty cries. Numb and heavy-winged with cold, the hen roused and shook herself. It took many minutes to free herself from the snow and to climb, half fluttering, half stumbling, out of the nest, deep-walled with white. Ookpik clucked to her and made the sounds of a cock owl bringing a lemming or a baby ptarmigan to the nest, but neither owl had had food since the blizzard began. The hen tried to fly but her heavy body flopped awkwardly in the snow for stiffness. When at last the slow circulation had crept back into her muscles, she rose into the air and the two owls floated over the place where the sanderlings crouched and out across the tundra.

As the snow fell on the still-warm eggs and the hard, bitter cold of the night gripped them, the life fires of the tiny embryos burned low. The crimson streams ran slower in the vessels that carried the racing blood from the food yolks to the embryos. After a time there slackened and finally ceased the furious activity of cells that grew and divided, grew again and divided to make owl bone and muscle and sinew. The pulsating red sacs under the great oversized heads hesitated, beat spasmodically, and were stilled. The six little owls-to-be were dead in the snow, and by their death, perhaps, hundreds of unborn lemmings and ptarmigans and Arctic hares had the greater chance of escaping death from the feathered ones that strike from the sky.

Farther up the ravine, several willow ptarmigans had been buried in a drift, where they had bedded for the night. The ptarmigans had flown over the ridge on the evening of the storm, dropping into the soft drifts so that never a print of their feet—clad in feathered snowshoes—was left to guide the foxes to their resting place. This was a rule of the game of life and death which the weak play with the strong. But tonight there was no need to observe the rules, for the snow would have obliterated all footprints and would have outwitted the keenest enemy—even as it drifted, by slow degrees, so deeply over the sleeping ptarmigans that they could not dig themselves out.

Five of the sanderling flock had died of the cold, and snow buntings by the score were stumbling and fluttering over the snow crust, too weak to stand when they tried to alight.

Now, with the passing of the storm, hunger was abroad on the great barrens. Most of the willows, food of the ptarmigans, were buried under snow. The dried heads of last year's weeds, which released their seeds to the snow buntings and the longspurs, wore glittering sheaths of ice. The lemmings, food of the foxes and the owls, were safe in their runways, and nowhere in this silent world was there food for shore birds that live on the shellfish and insects and other creatures of the water's edge. Now many hunters, both furred and feathered, were abroad during the night, the short, gray night of the Arctic spring. And when night wore into day the hunters still padded over the snow or beat on strong wings across the tundra, for the night's kill had not satisfied their hunger.

Among the hunters was Ookpik, the snowy owl. The coldest months of every winter, the icebound months, Ookpik spent hundreds of miles south of the barren grounds, where it was easier to find the little gray lemming mice that were his favorite food. During the storm nothing living had showed itself to Ookpik as he sailed over the plains and along the ridges that overlooked the sea, but today many small creatures moved over the tundra.

Along the east bank of the stream a flock of ptarmigans had found a few twigs of willow showing above the snow, part of a shrubby growth that had been as high as the antlers of a barren-grounds caribou until the snow had covered it. Now the ptarmigans could easily reach the topmost branches, and they nipped off the twigs in their bills, content with this food until the tender new buds of spring should be put forth. The flock still wore the white plumage of winter except for one or two of the cocks whose few brown feathers told of approaching summer and the mating season. When a ptarmigan in winter dress feeds on the snow fields, all of color about him is the black of bill and roving eye, and of the under tail feathers when he flies. Even his ancient enemies, the foxes and the owls, are deceived from a distance; but they, too, wear the Arctic's protective colorings.

Now Ookpik, as he came up the stream valley, saw among the willows the moving balls of shining black that were the ptarmigans' eyes. The white foe moved nearer, blending into the pale sky; the white prey moved, unfrightened, over the snow. There was a soft *whoosh* of wings—a scattering of feathers—and on the snow a red stain spread, red as a new-laid ptarmigan egg before the shell pigments have dried. Ookpik bore the ptarmigan in his talons over the ridge to the higher ground that was his lookout, where his mate awaited him. The two owls tore apart the warm flesh with their beaks, swallowing also the bones and feathers as was their custom, to cast them up later in neat pellets.

The gnawing pang of hunger was a sensation new to Silverbar. A week before, with the others of the sanderling flock, she had filled her stomach with shellfish gathered on the wide tidal flats of Hudson Bay. Days before that they had gorged on beach fleas on the coasts of New England, and on Hippa crabs on the sunny beaches of the south. In all the eight-thousand-mile journey northward from Patagonia there had been no lack of food.

The older sanderlings, patient in the acceptance of hardship, waited until the ebb tide, when they led Silverbar and the other year-old birds of the flock to the edge of the harbor ice. The beach was piled with irregular masses of ice and frozen spray, but the last tide had shifted the broken floe and on retreating had left a bare patch of mud flat. Already several hundred shore birds had gathered—all the early migrants from miles around who had escaped death in the snow. They were clustered so thickly that there was scarcely space for the sanderlings to alight, and every square inch of surface had been probed or dug by the bills of the waders. By deep probing in the stiff mud Silverbar found several shells coiled like snails, but they were empty. With Blackfoot and two of the yearling sanderlings, she flew up the beach for a mile, but snow carpeted the ground and the harbor ice and there was no food.

As the sanderlings hunted fruitlessly among the ice chunks, Tullugak the raven flew overhead and passed up the shore on deliberate wings.

Cr-r-r-uck! Cr-r-r-uck! he croaked hoarsely.

Tullugak had been patrolling the beach and the nearby tundra for miles, on the lookout for food. All the known carcasses which the ravens had resorted to for months had been covered by snow or carried away by the shifting of the bay ice. Now he had located the remains of a caribou which the wolves had run down and killed that morning, and he was calling the other ravens to the feast. Three jet-black birds, among them Tullugak's mate, were walking briskly about over the bay ice hunting a whale carcass. The whale had come ashore months before, providing almost a winter's supply of food for Tullugak and his kin, who lived the year round in the vicinity of the bay. Now the storm had opened a channel into which the shifting ice masses had pushed the dead whale and closed over it. At the welcome food cry of Tullugak, the three ravens sprang into the air and followed him across the tundra to pick off the few shreds of meat that remained on the bones of the caribou.

The next night the wind shifted and the thaw began.

Day by day the blanket of snow grew thinner. Irregular holes rent the white covering—brown holes where the naked earth showed through, green holes where ponds that still held their hearts of ice were uncovered. Hillside trickles grew to rivulets and rivulets to rushing torrents as the Arctic sent its melted snows to the sea, to eat jagged cuts and gullies through the salt ice, to accumulate in pools along the shore. Lakes brimmed with the clear cold water and teemed with new life as the young of crane flies and May flies stirred in the bottom muds and the larvae of the northland's myriad mosquitoes squirmed in the water.

As the drifts melted away and the low-lying grasslands became flooded, the lemming burrows, which honeycombed the Arctic underworld with hundreds of miles of tunnels, became uninhabitable. The quiet runways, the peaceful grass-lined burrows that had been secure from even the fiercest blizzards of winter, now knew the terrors of rushing waters, of swirling floods. As many of the lemmings as could escape took refuge on high rocks and gravel ridges and sunned their plump gray bodies, quickly forgetting the dark horror from which they had lately fled.

Now hundreds of migrants arrived from the south each day and the tundra heard other noises besides the booming cries of the cock owls and the bark of the foxes. There were the voices of curlews and plovers and knots, of terns and gulls and ducks from the south. There were the braying cries of the stilt sandpipers and the tinkling song of the redbacks; there was the shrill bubbling of the Baird sandpiper, akin to the sleigh-bell chorus of spring peepers in the smoky twilight of a New England spring.

As the patches of earth spread over the snow fields, the sanderlings, plovers, and turnstones gathered in the cleared spots, finding abundant food. Only the knots resorted to the unthawed marshes and the protected hollows of the plains, where sedges and weeds lifted dry seed heads above the snow and rattled when the wind blew and dropped their seed for the birds.

Most of the sanderlings and the knots passed on to the distant islands scattered far over the Arctic sea, where they made their nests and brought forth their young. But Silverbar and Blackfoot and others of the sanderlings remained near the bay shaped like a leaping porpoise, along with turnstones, plovers, and many other shore birds. Hundreds of terns were preparing to nest on nearby islands, where they would be safe from the foxes, while most of the gulls retired inland to the shores of the small lakes which dotted the Arctic plains in summer.

In time Silverbar accepted Blackfoot as her mate and the pair withdrew to a stony plateau overlooking the sea. The rocks were clothed with mosses and soft gray lichens, first of all plants to cover the open and wind-swept places of the earth. There was a sparse growth of dwarf willow, with bursting leaf buds and ripe catkins. From scattered clumps of green the flowers of the wild betony lifted white faces to the sun, and over the south slope of the hill was a pool fed by melting snow and draining to the sea by way of an old streambed.

Now Blackfoot grew more aggressive and fought bitterly with every cock who infringed upon his chosen territory. After such a combat he paraded before Silverbar, ruffling his feathers. While she watched in silence he leaped into the air and hovered on fluttering wings, uttering neighing cries. This he did most often in the evening as the shadows lay purple on the eastern slopes of the hills.

On the edge of a clump of betony Silverbar prepared the nest, a shallow depression which she molded to her body by turning round and round. She lined the bottom with last year's dried leaves from a willow that grew prostrate along the ground, bringing the leaves one at a time and arranging them in the nest along with some bits of lichen. Soon four eggs lay on the willow leaves, and now Silverbar began the long vigil during which she must keep all wild things of the tundra from discovering the place of her nest.

During her first night alone with the four eggs Silverbar heard a sound new to the tundra that year, a harsh scream that came again and again out of the shadows. At early dawn light she saw two birds, dark of body and wing, flying low over the tundra. The newcomers were jaegers, birds of the gull tribe turned hawk to rob and kill. From that time on, the cries, like weird laughter, rang every night on the barrens.

More and more jaegers arrived each day, some from the fishing grounds of the North Atlantic, where they had lived by stealing fish from the gulls and shearwaters, others from the warm oceans of half the globe. Now the jaegers became the scourge of all the tundra. Singly or in twos and threes they beat back and forth over the open places, on the watch for a solitary sandpiper or plover or phalarope who, being defenseless, would provide easy game. They whirled down in sudden attack on the flocks of shore birds feeding on the broad, weed-strewn mud flats, hoping to separate a single bird from its fellows for the swift pursuit that ends in death. They harried the gulls on the bay, tormenting them until they disgorged the fish they had caught. They hunted among the rock crevices and little mounds of stones, where they often surprised a lemming sunning himself at the mouth of his burrow or came upon a snow bunting brooding her eggs. They perched on rocky elevations or ridges from which they watched the pattern of the rolling tundra, mottled dark and light with moss and gravel, with lichen and shale. Even the fierce eyes of a jaeger could not distinguish at a distance the speckled eggs of the many birds that lay unconcealed on the open plain. So skillful was the camouflage of the tundra that only by sudden movement did a nesting bird or a foraging lemming betray its presence.

Now for twenty hours out of the twenty-four the tundra lay in sun and for four hours it slept in soft twilight. Arctic willow and saxifrage, wild betony and crowberry hastened to put forth new leaves to draw to themselves the strength of the sun. Into a few brief and sun-filled weeks the plants of the Arctic must crowd a lifetime of living. Only the kernel of life, fortified and protected, endures the months of darkness and cold.

Soon the cloak of the tundra was embroidered with many flowers: first, the white cups of the mountain avens; then the purple of saxifrage; then the yellow of the buttercup glades, loud with the drone of bees trampling the shiny golden petals and jostling the laden anthers, so that each bore away its load of pollen on the bristles of its body. The tundra was gay, too, with moving bits of color, for the midday sun coaxed out butterflies from the willow thickets where they drooped and hid when the colder airs blew or when clouds stole between earth and sun.

In temperate lands the birds sing their sweetest songs in the dim light that falls after sundown and comes before dawn. But in the Arctic barrens the June sun dips so briefly below the horizon that each of the night hours is an hour of twilight or song light, filled with the bubbling song of the longspur and the calls of the horned lark.

On a day in June a pair of phalaropes drifted, light as corks, on the glazed sheet of the sanderlings' pool. Now and then they spun in a circle by rapid thrusts of their lobed feet, then jabbed again and again with needlelike bills to capture the insects stirred up by their movements. The phalaropes had spent the winter on the open sea far to the southward, following the whales and the ever-drifting clouds of whale food. On their migration they had come northward by an ocean route as far as possible before striking inland. The phalaropes prepared a nest on the south slope of the ridge not far from the sanderling nest and lined it, as most of the tundra nests are lined, with willow leaves and catkins. Then the cock phalarope took charge of the nest, to sit for eighteen days, warming the eggs to life.

By day the soft *coo-a-bee, coo-a-bee* of the knots came down flutelike from the hills, where on the plateaus the nests lay hidden amid the curling brown tufts of Arctic sedges and the leaves of the mountain avens. Every evening Silverbar watched a solitary knot tumbling and soaring in the still air over the low mounds of the hills. The song of Canutus, the knot, was heard by other knots along miles of hilltop and by the turnstones and sandpipers on the tide flats of the bay. But another heard and responded to it most of all, his small dappled mate, who was brooding their four eggs in the nest far below.

Then for a season many of the voices of the tundra were hushed, as all over the barren eggs were hatching and there were young to be fed and concealed from enemies.

When Silverbar had begun to brood her eggs, the moon had been at the full. Since then it had dwindled to a thin white rim in the sky and now had grown again to the quarter, so that once more the tides in the bay were slacker and milder. One morning when the shore birds gathered over the flats to feed on the ebb tide, Silverbar did not join them. Throughout the night there had been sounds in the eggs under her breast feathers, now worn and frayed. They were the peckings of the sanderling chicks, after twenty-three days made ready for life. Silverbar inclined her head and listened to the sounds; sometimes she withdrew a little from the eggs and watched them intently.

On the nearby ridge a Lapland longspur was singing his tinkling, many-syllabled song, mounting high into the air again and again and spilling out his song as he lowered himself on widespread wings to the grass. The little bird had a feather-lined nest on the edge of the phalarope pool, where his mate was brooding their six eggs. The longspur was glad of the brightness and warmth of noonday and unmindful of the shadow that dropped between him and the sun as Kigavik, the gyrfalcon, fell from the sky. Silverbar neither heard the song of the longspur nor was conscious of its sudden ceasing; nor did she notice when a single breast feather fluttered downward almost at her side. She was watching a hole that had appeared in one of the eggs. The only sound she heard was a thin, mouse-like squeak, the first cry of her young. By the time the gyrfalcon had reached his eyrie on a crag of rock facing northward to the sea and had fed the longspur to his nestlings, the first sanderling chick had emerged from its shell and two more of the eggs were cracked.

Now for the first time an abiding fear entered the heart of Silverbar—the fear of all wild things for the safety of their helpless young. With quickened senses she perceived the life of the tundra—with ears sharpened to hear the screams of the jaegers harrying the shore birds on the tide flats—with eyes quickened to note the white flicker of a gyrfalcon's wing.

After the fourth chick had hatched, Silverbar began to carry the shells, piece by piece, away from the nest. So countless generations of sanderlings had done before her, by their cunning outwitting the ravens and foxes. Not even the sharp-eyed falcon from his rock perch nor the jaegers watching for lemmings to come out of their holes saw the movement of the little brown-mottled bird as she worked her way, with infinite stealth, among the clumps of betony or pressed her body closely to the wiry tundra grass. Only the eyes of the lemmings who ran in and out among the sedges or sunned themselves on flat rocks near their burrows saw the mother sanderling until she reached the bottom of the ravine on the far side of the ridge. But the lemmings were gentle creatures who neither feared nor were feared by the sanderling.

All through the brief night that followed the hatching of the fourth chick Silverbar worked, and when the sun had come around to the east again she was hiding the last shell in the gravel of the ravine. A polar fox passed near her, making no sound as he trotted with sure foot over the shales. His eye gleamed as he watched the mother bird, and he sniffed the air, believing she had young nearby. Silverbar flew to the willows farther up the ravine and watched the fox uncover the shells and nose them. As he started up the slope of the ravine the sanderling fluttered toward him, tumbling to the ground as though hurt, flapping her wings, creeping over the gravel. All the while she uttered a high-pitched note like the cry of her own young. The fox rushed at her. Silverbar rose rapidly into the air and flew over the crest of the ridge, only to reappear from another quarter, tantalizing the fox into following her. So by degrees she led him over the ridge and southward into a marshy bottom fed by the overflow of upland streams.

As the fox trotted up the slope, the cock phalarope on the nest heard a low *Plip! Plip! Chiss-ick! Chiss-ick!* from the hen, who was on guard nearby and had seen the fox coming up the slope. The cock crept silently from the nest, through the grassy tunnels he had fashioned as runways of escape, until he came to the waterside where his mate was awaiting him. The two birds sailed into the middle of the pool and swam anxiously in circles, preening their feathers, jabbing long bills into the water in a pretense of feeding until the air came clean again, untainted by the musky smell of fox. The cock's breast showed a worn spot where the feathers had frayed away, for the phalarope chicks were soon to hatch.

When Silverbar had led the fox far enough from her young she circled around by the bay flats, pausing to feed nervously for a few minutes at the edge of the salty tide. Then she flew swiftly to the betony clump and the four chicks on which the down was yet dark with the dampness of the egg, although soon it would dry to tones of buff and sand and chestnut.

Now the sanderling mother knew by instinct that the depression in the tundra, lined with dry leaves and lichens and molded to the shape of her breast, was no longer a safe place for her young. The gleaming eyes of the fox—the soft pad, pad of his feet on the shales—the twitch of his nostrils testing the air for scent of her chicks—became for her the symbols of a thousand dangers, formless and without name.

When the sun had rolled so low on the horizon that only the high cliff with the eyrie of the gyrfalcon caught and reflected its gleam, Silverbar led the four chicks away into the vast grayness of the tundra.

Throughout the long days the sanderling with her chicks wandered over the stony plains, gathering the young ones under her during the short chill nights or when sudden gusts of rain drove across the barrens. She led them by the shores of brimming fresh-water lakes into which loons dropped on whistling wings to feed their young. Strange new food was to be found on the shores of the lakes and in the swelling turbulence of feeder streams. The young sanderlings learned to catch insects or to find their larvae in the streams. They learned, too, to press themselves flat against the ground when they heard their mother's danger cry and to lie quite still among the stones until her signal brought them crowding about her with fine, high-pitched squeakings. So they escaped the jaegers, the owls, and the foxes.

By the seventh day after hatching, the chicks had quill feathers a third grown on their wings, although their bodies were still covered with down. After four more suns the wings and shoulders were fully clothed in feathers, and when they were two weeks old the fledgling sanderlings could fly with their mother from lake to lake.

Now the sun dipped farther below the horizon; the grayness of the nights deepened; the hours of twilight lengthened. The rains that came more often and lashed with sharper violence were matched by a gentler rain as the flowers of the tundra dropped their petals. The foodstuffs—the starches and the fats—had been stored away in the seeds to nourish the precious embryos, into which had passed the immortal substance of the parent plants. The summer's work was done. No more need of bright petals to lure the pollen-carrying bees; so cast them off. No more need of leaves spread to catch the sunshine and harness it to chlorophyll and air and water. Let the green pigments fade. Put on the reds and yellows, then let the leaves fall, too, and the stalks wither away. Summer is dying.

Soon the first white hairs appeared in the coats of the weasels, and the hair of the caribou began to lengthen. Many of the cock sanderlings, who had been gathering in flocks about the fresh-water lakes almost from the time the chicks had begun to hatch, had already left for the south. Among them was Blackfoot. On the mud flats of the bay young sandpipers gathered by the thousand and in the new-found joy of flight their flocks soared and swooped over a calm sea. The knots had brought their young down from the hills to the seacoast, and day by day more of the adults were leaving. On the pool near the place where Silverbar had brooded her eggs, a family of three young phalaropes now spun with lobed feet and jabbed for insects along the shore. The cock and the hen phalarope were already hundreds of miles to the east, setting a course southward over open ocean.

There came a day in August when Silverbar, who had been feeding with her grown young on the shores of the bay in company with other sanderlings, suddenly rose into the air with some twoscore of the older birds. The little flock wheeled out over the bay in a wide circle, flashing white wing bars; they returned, crying loudly as they passed over the flats where the young were still running and probing at the edge of the curling wavelets; they turned their heads to the south and were gone.

There was no need for the parent birds to remain longer in the Arctic. The nesting was done; the eggs had been faithfully brooded; the young had been taught to find food, to hide from enemies, to know the rules of the game of life and death. Later, when they were strong for the journey down the coastlines of two continents, the young birds would follow, finding the way by inherited memory. Meanwhile the older sanderlings felt the call of the warm south; they would follow the sun.

That evening about sunset Silverbar's four young, now wandering with a score of other fledgling sanderlings, came to an inland plain cut off from the sea by a coastwise ridge and rimmed to the south by higher hills. The floor of the plain was grassy and patched in many places with the softer, intense green of marsh. The sanderlings came into the plain along a meandering stream and settled on its banks for the night.

To the sanderlings' ears all the plain was alive with a kind of rustling—a soft murmur—a persistent stirring. It was like the sound of the wind when it moves through pine trees; but on the great barrens there are no trees. It was like the soft spilling of a stream over its bed, water striking stone, pebble rubbing against pebble. But tonight the stream was locked beneath the first thin ice of the summer's end.

The sound was the stirring of many wings, the passage of many feathered bodies through the low vegetation of the plain, the murmur of myriad bird voices. The flocks of the golden plover were gathering. From the wide beaches of the sea, from the shores of the bay shaped like a leaping porpoise, from all the tundras and uplands for miles around, the black-bellied birds with the golden-speckled backs were assembling on the plain.

The plovers were in a state of excitement that mounted as evening shadows cloaked the tundra and darkness spread over the Arctic world, save for a fiery glow on the horizon, as though the wind stirred the ashes of the sun's fires. The sound of the bird voices, constantly augmented by new arrivals and increasing in volume as the mass excitement grew, swept over the plain like a wind. Above the general murmur there arose at intervals the high, quavering cries of the leaders of the flock.

About midnight the flight began. The first flock of some threescore birds rose into the air, circled over the plain, and straightening out into flight formation headed south and east. Another and another flock found its wings and hurtled after the leaders, flying low over the tundra that rolled like a deep purple sea beneath them. There was strength and grace and beauty in every stroke of the pointed wings; there was power without end for the journey.

Que-e-e-e-ah! Que-e-e-e-ah!

High-pitched and quavering, the calls of the migrants came down clear from the sky.

Que-e-e-e-ah! Que-e-e-e-ah!

Every bird of the tundra heard the call and stirred in vague unrest at its urgency.

Among those who heard there must have been the young plovers, the birds of the year, scattered in little wandering groups over the tundra. But none among them joined the flight of the older birds. Not until weeks later, alone and with none to guide them, would they undertake the journey.

From the end of the first hour onward the flight was no longer divided into flocks but became continuous. Now a mighty river of birds poured through the sky, lengthening as it flowed south and east across the barrens, across the head of the northland bay, and on and on through skies that lightened to the coming of another day.

People said of it that it was the greatest golden plover flight of many years. Father Nicollet, the old priest in his mission on the west shore of Hudson Bay, declared that it reminded him of the great flights he had seen in his youth, before the gunners had thinned the plover flocks to a remnant of their former size. Eskimos and trappers and traders along the Bay raised their eyes to the morning sky to watch the last of the flight crossing the Bay and fading into the east.

Somewhere in the mists beyond them lay the rocky shores of Labrador, carpeted with the bushes of the crowberry hung with purple fruit; beyond lay the tide flats of Nova Scotia. From Labrador to Nova Scotia the birds would slowly work their way, feeding on the ripening crowberries, on beetles and caterpillars and shellfish, growing fat and storing away energy to be burned by active muscles.

But soon there would come a day when again the flocks would spring into the air, this time to head southward into the misty horizon where sky met sea. Southward they would lay their course across more than two thousand miles of ocean from Nova Scotia to South America. They would be seen by men in boats far at sea, flying a swift, straight course low to the water, like those who know their destination and suffer nothing to deter them.

Some, perhaps, would fall by the way. Some, old or sick, would drop out of the caravan and creep away into a solitary place to die; others would be picked off by gunners, defying the law for the fancied pleasure of stopping in full flight a brave and fiercely burning life; still others, perhaps, would fall in exhaustion into the sea. But no awareness of possible failure or disaster dwelt in the moving host, flying with sweet pipings through the northern sky. In them burned once more the fever of migration, consuming with its fires all other desires and passions.

4

SUMMER'S END

It was September before the sanderlings, now in whitening plumage, ran again on the island beach or hunted Hippa crabs in the ebbing tide at the point of land called Ship's Shoal. Their flight from the northern tundras had been broken by many feeding stops on the wide mud flats of Hudson Bay and James Bay and on the ocean beaches from New England southward. In their fall migration the birds were unhurried, the racial urge that drove them northward in the spring having been satisfied. As the winds and the sun dictated, they drifted southward, their flocks now growing as more birds from the north joined them, now dwindling as more and more of the migrants found their customary winter home and dropped behind. Only the fringe of the great southward wave of shore birds would push on and on to the southernmost part of South America.

As the cries of the returning shore birds rose once more from the frothy edge of the surf and the whistle of the curlews sounded again in the salt marshes, there were other signs of the summer's end. By September the eels of the sound country had begun to drop downstream to the sea. The eels came down from the hills and the upland grasslands. They came from cypress swamps where black-watered rivers had their beginnings; they moved across the tidal plain that dropped in six giant steps to the sea. In the river estuaries and in the sounds they joined their mates-to-be. Soon, in silvery wedding dress, they would follow the ebbing tides to the sea, to find—and lose—themselves in the black abysses of mid-ocean.

By September, the young shad, come from the eggs shed in river and stream by the spawning runs of spring, were moving with the river water to the sea. At first they moved slowly in the vaster currents as the sluggish rivers broadened toward their estuaries. Soon, however, the speed of the little fish, no longer than a man's finger, would quicken when the fall rains came and the wind changed, chilling the water and driving the fish to the warmer sea.

By September the last of the season's hatch of young shrimp were coming into the sounds through the inlets from the open sea. The coming of the young was symbolic of another journey which no man had seen and no man could describe—a journey taken weeks before by the elder generation of shrimp. All through the spring and summer more and more of the grown shrimp, come to maturity at the age of a year, had been slipping away from the coastal waters, journeying out across the continental shelf, descending the blue slopes of undersea valleys. From this journey they never returned, but their young, after several weeks of ocean life, were brought by the sea into the protected inside waters. All through the summer and fall the baby shrimp were brought into the sounds and river mouths—seeking warm shallows where brackish water lay over muddy bottoms. Here they fed eagerly on the abundant food and found shelter from hungry fish in the carpeting eel grass. And as they grew rapidly, the young turned once more to the sea, seeking its bitter waters and its deeper rhythms. Even as the youngest shrimp from the last spawning of the season came through the inlets on each flood tide of September, the larger young were moving out through the sounds to the sea.

By September the panicles of the sea oats in the dunes had turned a golden brown. As the marshes lay under the sun, they glowed with the soft greens and browns of the salt meadow grass, the warm purples of the rushes, and the scarlet of the marsh samphire. Already the gum trees were like red flares set in the swamps of the riverbanks. The tang of autumn was in the night air, and as it rolled over the warmer marshes it turned to mist, hiding the herons who stood among the grasses at dawn; hiding from the eyes of the hawks the meadow mice who ran along the paths they had made through the marshes by the patient felling of thousands of marsh-grass stems; hiding the schools of silversides in the sound from the terns who fluttered above the rolling white sea, and caught no fish until the sun had cleared away the mists.

The chill night air brought a restlessness to many fish scattered widely throughout the sound. They were steely gray fish with large scales and a low, four-spined fin set on the back like a spread sail. The fish were mullet who had lived throughout the summer in the sound and estuary, roving solitary among the eel grass and widgeon grass, feeding on the litter of animal and vegetable fragments of the bottom mud. But every fall the mullet left the sounds and made a far sea journey, in the course of which they brought forth the next generation. And so the first chill of fall stirred in the fish the feeling of the sea's rhythm and awakened the instinct of migration.

The chilling waters and the tidal cycles of the summer's end brought to many of the young fish of the sound country, also, a summons to return to the sea. Among these were the young pompano and mullet, silversides and killifish, who lived in the pond called Mullet Pond, where the dunes of the barrier island fell away to the flat sands of the Ship's Shoal. These young fish had been spawned in the sea, but had found their way to the pond through a temporary cut earlier that year.

On a day when the full harvest moon sailed like a white balloon in the sky, the tides, which had grown in strength as the moon swelled to roundness, began to wash out a gully across the inlet beach. Only on the highest tides did the torpid pond receive water from the ocean. Now the beat of the waves and the strong backwash that sucked away the loose sand had found the weak place in the beach, where a cut had been made before, and in less time than it took a fishing launch to cross from the mainland docks to the banks a narrow gully or slough had been cut through to the pond. Not more than a dozen feet across, it made a bottleneck into which the surf rolled as the waves broke on the beach. The water surged and seethed as in a mill race, hissing and foaming. Wave after wave poured through the slough and into the pond. They dug out an uneven, corrugated bottom over which the water leaped and tobogganed. They spread out into the marshes that backed the pond, seeping silently and stealthily among the grass stems and the reddening stalks of the marsh samphire. Into the marshes they carried the frothy brown scud thrown off by the waves. The sandy foam filled the spaces between the grass stalks so closely that the marsh looked like a beach thickly grown with short grass; in reality the grass stood a foot in water and only the upper third of the stalks showed above the froth.

Leaping and racing, foaming and swirling, the incoming flood brought release to the myriads of small fishes that had been imprisoned in the pond. Now in thousands they poured out of the pond and out of the marshes. They raced in mad confusion to meet the clean, cold water. In their excitement they let the flood take them, toss them, turn them over and over. Reaching midchannel of the slough they leaped high in the air again and again, sparkling bits of animate silver, like a swarm of glittering insects that rose and fell, rose again and fell. There the water seized them and held them back in their wild dash to the sea, so that many of them were caught on the slopes of the waves and held, tails uppermost, struggling helplessly against the might of the water. When finally the waves released them they raced down the slough to the ocean, where they knew once more the rolling breakers, the clean sandy bottoms, the cool green waters.

How did the pond and the marshes hold them all? On they came, in school after school, flashing bright among the marsh grasses, leaping and bounding out of the pond. For more than an hour the exodus continued, with scarcely a break in the hurrying schools. Perhaps they had come in, many of them, on the last spring tide when the moon was a pencil stroke of silver in the sky. And now the moon had grown fat and round and another spring tide, a rollicking, roistering, rough-and-ready tide, called them back to the sea again.

On they went, passing through the surf line where the white-capped waves were tumbling. On they went, most of them, past the smoother green swells to the second line of surf, where shoals tripped the waves coming in from the open sea and sent them sprawling in white confusion. But there were terns fishing above the surf, and thousands of the small migrants went no farther than the portals of the sea.

Now there came days when the sky was gray as a mullet's back, with clouds like the flung spray of waves. The wind, that throughout most of the summer had blown from the southwest, began to veer toward the north. On such mornings large mullet could be seen jumping in the estuary and over the shoals of the sound. On the ocean beaches fishermen's boats were drawn up on the sand. Gray piles of netting lay in the boats. Men stood on the beach, with eyes on the water, patiently waiting. The fishermen knew that mullet were gathering in schools throughout the sound because of the change in the weather. They knew that soon the schools would run out through the inlet before the wind and then would pass down along the coast, keeping, as the fishermen had told it from one generation to the next, "their right eyes to the beach." Other mullet would come down from the sounds that lay to the north and still others would come by the outside passage, following down along the chain of barrier islands. So the fishermen waited, confident in their generations of tested lore; and the boats waited with the nets that were empty of fish.

Other fishers besides the men awaited the runs of mullet. Among them was Pandion, the fish hawk, whom the mullet fishermen watched every day as he floated, a small dark cloud, in wide circles in the sky. To pass the hours as they stood watch on the sound beach or among the dunes, the fishermen wagered among themselves when the osprey would dive.

Pandion had a nest in a clump of loblolly pines on the shore of the river three miles away. There he and his mate had hatched and reared a brood of three young that season. At first the young had been clothed in down that was the color of old, decaying tree stumps; now they had grown their pinions and had gone away to fish for themselves, but Pandion and his mate, who had been faithful to each other throughout life, continued to live in the nest which they had used year after year.

The nest was six feet across at its base and more than half as wide at the top. Its bulk would have overflowed any of the farm carts that were drawn by mules along the dirt roads of the

sound country. The two ospreys had repaired the nest and added to it during the years anything they could find washed up on the beaches by the tides. Now practically the whole top of a forty-foot pine served as support for the nest, and the great weight of sticks, branches, and pieces of sod had killed all but a few of the lower branches. In the course of years the ospreys had woven or worked into the nest a twenty-foot piece of haul seine with ropes attached that they had picked up on the shore of the sound, perhaps a dozen cork floats from fishing gear, many cockle and oyster shells, part of the skeleton of an eagle, parchmentlike strings of the egg cases of conchs, a broken oar, part of a fisherman's boot, tangled mats of seaweed.

In the lower layers of the huge, decaying mass many small birds had found nesting places. That summer there had been three families of sparrows, four of starlings, and one of the Carolina wren. In the spring an owl had taken up quarters in the osprey nest, and once there had been a green heron. All these lodgers Pandion had suffered good-naturedly.

After the third day of grayness and chill, the sun broke through the clouds. Watched by the mullet fishermen, Pandion sailed on set wings, riding the mounting columns of warm air that shimmered upward from the water. Far below him the water was like green silk rippling in a breeze. The terns and skimmers resting on the shoals of the sound were the size of robins. The black, glistening backs of a school of dolphins, diving and rolling, moved, a dark serpent, over the face of the sound. The amber eyes of Pandion flickered as a whipper ray leaped three times from the water, coming down with a sharp spat that was carried away on the wind and lost.

A shadow took form on the green screen beneath the osprey and the surface dimpled as a fish nosed at the film. In the sound two hundred feet below the fish hawk, Mugil, the mullet—the leaper—gathered his strength and flung himself in exhilaration into the air. As he was flexing his muscles for a third leap a dark form fell out of the sky and viselike talons seized him. The mullet weighed more than a pound, but Pandion carried him easily in his taloned feet, bearing the fish across the sound and to the nest three miles away.

Flying up the river from the estuary the osprey carried the mullet headfirst in his talons. As he neared the nest he relaxed his grip with the left foot and, checking flight, alighted on the outer branches of the nest with the fish still gripped in his right foot. Pandion lingered over his meal of fish for more than an hour, and when his mate came near he crouched low over the mullet and hissed at her. Now that the nesting was done, every bird must fish for itself.

Later in the day, as he returned down the river to fish, Pandion swooped low to the water and for the space of a dozen wing beats dragged his feet in the river, cleansing them of the adhering fish slime.

On his return Pandion was watched by the sharp eyes of a large brown bird perched in one of the pines on the west bank of the river, overlooking the marshes of the estuary. White Tip, the bald eagle, lived as a pirate, never fishing for himself when he could steal from the ospreys of the surrounding country. When Pandion moved out over the sound the eagle followed, mounting into the air and taking up a position far above the fish hawk.

For an hour two dark forms circled in the sky. Then from his high station White Tip saw the body of the osprey suddenly dwindle to sparrow size as he fell in a straight drop, saw the white spray mount from the water as the fish hawk disappeared. After the passing of thirty seconds Pandion emerged from the water, mounting straight for fifty feet with short, heavy wing beats and then leveling out into straight flight toward the river's mouth.

Watching him, White Tip knew that the osprey had caught a fish and was taking it home to the nest in the pines. With a shrill scream that fell down through the sky to the ears of the osprey, the eagle whirled in pursuit, keeping his elevation of a thousand feet above the fish hawk.

Pandion cried out in annoyance and alarm, redoubling the force of his wing beats in an effort to reach the cover of the pines before his tormentor should attack. The speed of the hawk was retarded by the weight of the catfish that he carried and by the convulsive struggles of the fish, held firmly in the strong talons.

Between the island and the mainland and several minutes' flight from the mouth of the river the eagle gained a position directly over the hawk. On half-closed wings he dropped with terrific speed. The wind whined through his feathers. As he passed the osprey he whirled in air, back to the water, presenting his talons to the attack. Pandion dodged and twisted, eluding the eight curved scimitars. Before White Tip could recover himself Pandion had shot aloft two hundred, five hundred feet. The eagle hurtled after him, mounted above him. But even as he began the stoop, the fish hawk, in another upward soaring, surmounted the position of his enemy.

Meanwhile the fish, drained of life by separation from the water, grew limp as all its struggles ceased. Like a mist gathering on a clear glass surface, a film clouded its eyes. Soon the iridescent greens and golds that made its body, in life, a thing of beauty had faded to dullness.

By turns rising and swooping, hawk and eagle rose to a great height, into the empty places of the air, of which the sound and its shoals and white sands had no part.

Cheep! Cheep! Chezeek! Chezeek! screamed Pandion in a frenzy of excitement.

A dozen white feathers, ripped from his breast as he barely evaded White Tip's talons on the last stoop, fluttered earthward. Of a sudden the osprey bent his wings sharply and dropped like a stone toward the water. The wind roared in his ears, half blinded him, plucked at his feathers as the sound rushed up to meet him. It was his final effort to outwit a stronger and more enduring enemy. But from above, the relentless dark form fell even faster than Pandion, gained on him, passed him as the fishing boats on the sound grew big as gulls afloat, whirled and tore the fish from his grasp.

The eagle carried the fish to his pine-tree perch to rend it, muscle from bone. By the time he reached the perch Pandion was beating out heavily over the inlet to new fishing grounds at sea.

5
WINDS BLOWING SEAWARD

The next morning the north wind was tearing the crests off the waves as they came over the inlet bar, so that each was trailing a heavy smoke of spray. Mullet were jumping in the channel, excited by the change in the wind. In the shallow river estuary and over the many shoals of the sound, the fish sensed the sudden chill that passed to the water from the air moving over it. The mullet began to seek the deeper waters which held the stored warmth of the sun. Now from all parts of the sound they were assembling in large schools that moved toward the channels of the sound. The channels led to the inlet, and the inlet was the gateway to the open sea.

The wind blew from the north. It blew down the river, and the fish moved before it to the estuary. It blew across the sound to the inlet, and the fish ran before it to the sea.

The ebbing tide carried the mullet through the deeper green glooms and over the white sandy bottom of the channel, scoured clean of living things by the strong currents that raced through it each day, twice running seaward, twice landward. Above them, as they moved, the surface of the water was broken into a thousand glittering facets that shone with the sun's gold. One after another the mullet rose to the shimmering ceiling of the sound. One after another they flexed their bodies in a quickening rhythm, gathering their strength and leaping into the air.

Going out with the tide the mullet passed a long, narrow sandspit called Herring Gull Shoal, where a wall of massive stone was built along the channel to prevent the washing in of the loose sand. Green, turgid fronds of seaweed were anchored by their holdfasts to the stones, which were crusted whitely with barnacles and oysters. From the shadow of one of the stones of the breakwater a pair of small, malignant eyes watched the mullet as they passed seaward. The eyes belonged to the fifteen-pound conger eel who lived among the rocks. The thick-bodied conger preyed on the schools of fish that roved down along the dark wall of the breakwater, hurling itself out of its gloomy cavern to seize them in its jaws.

In the upper layer of water, a dozen feet above the swimming mullet, schools of silversides quivered in formation, each fishlet a gleaming mote reflecting the sunlight. From time to time scores of them leaped out of the water, bursting through the surface film of the fish's world and falling back again like raindrops—first denting, then piercing the tough skin between air and water.

Past a dozen sandspits of the sound, each with its little colony of resting gulls, the tide took the mullet. On an old shell rock which the sea was in process of turning into an island by dropping silt and sand among the shells and by bringing, on its ebb tides, the seeds of marsh grasses to bind the soil, two gulls were hunting busily for sunray clams, which lay half buried in the wet sand. Finding them, the gulls chipped away at the heavy, vitreous shells, rayed with bands of fawn color and lilac. After much work with their strong bills the gulls were able to crack the shells and eat the soft clam bodies within.

On the mullet went, past the big inlet buoy that was leaning toward the sea with the press of the tide. Its iron bulk rose and fell with the water, even as the music of its iron throat changed pitch and tempo with the changing rhythms of the sea. The inlet buoy was a cosmos unto itself, rolling in the waters of the sound. Ebb tide and flood tide were of its own making, coming alternately as the buoy lifted to the passing of a wave and rolled in its trough.

The buoy had not been taken in for scraping and repainting since the previous spring, and it was thickly crusted with the shells of barnacles and mussels and with saclike sea squirts and the soft moss patches of the bryozoa. Deposits of sand and silt and green threads of algae had lodged in the many crevices between the shells and among the rootlike attachments of the dense mat of animals. Over and among this thick, living growth, slender-bodied animals called amphipods, in jointed armor, clambered in and out in endless search of food; starfish crept over the oysters and mussels and preyed upon them, gripping the shells with the sucking discs of their strong arms and forcing them open. Among the shells the flowerlets of the sea anemones opened and closed, spreading fleshy tentacles to seize food from the water. Most of the twenty or more kinds of sea animals that lived on the buoy had come to it months before, during the season when the waters of the sound and inlet swarmed with larvae. Many of these myriad beings, as transparent as glass and more fragile, were doomed to die in infancy unless they found a solid place of attachment. Those that chanced upon the great bulk of the buoy in the sound attached themselves by cementing fluids from their own bodies or by byssus thread or holdfast. There they would remain throughout life, a part of the swaying world, rolling in watery space.

Within the inlet the channel widened and the pale-green water grew murky with the wave scourings of loose sand. On the mullet went. The mutter and rumble of the surf grew. With their sensitive flanks the fish perceived the heavy jar and thud of sea vibrations. The changing pulse of the sea was caused by the long inlet bar, where the water foamed to a white froth as the waves spilled over it. Now the mullet passed out through the channel and felt the longer rhythms of the sea—the rise, the sudden lift and fall of waves come from the deep Atlantic. Just outside the first surf line the mullet leaped in these larger swells of ocean. One after another swam upward to the surface and jumped into the air, falling back with a white splash to resume its place in the moving school.

The lookout who stood on a high dune above the inlet saw the first of the mullet running out of the sound. With practiced eye, he estimated the size and speed of the school from the spurts of spray when the mullet jumped. Although three boats with their crews were waiting farther down the ocean beach, he gave no signal at the passing of the first mullet. The tide was still on the ebb; the pull of the water was seaward and the nets could not be drawn against it.

The dunes are a place of high winds and driven sand, of salt spray and sun. Now the wind is from the north. In the hollows of the dunes the beach grasses lean in the wind and with their pointed tips write endless circles in the sand. From the barrier beach the wind is picking up the loose sand and carrying it seaward in a haze of white. From a distance the air above the banks looks murky, as though a light mist is rising from the ground.

The fishermen on the banks do not see the sand haze; they feel its sting in eyes and face; they feel it as it sifts into their hair and through their clothing. They take out their handkerchiefs and tie them across their faces, and they pull long-visored caps low on their heads. A wind from the north means sand in your face and rough seas under your boat keel, but it means mullet, too.

The sun is hot as it beats down on the men standing on the beach. Some of the women and children are there, too, to help their men with the ropes. The children are bare-footed, wading in the pools left in the scoured-out depressions of the beach, ribbed with sand waves.

The tide has turned, and now one of the boats is shot out between the breakers to be ready for the fish when they come. It's not easy, launching a boat in this surf. The men leap to their places like parts of a machine. The boat rights itself, wallows into the green swells. Just outside the surf line the men wait at the oars. The captain stands in the bow, arms folded, leg muscles flexing to the rise and fall of the boat, his eyes on the water, looking toward the inlet.

Somewhere in that green water there are fish—hundreds of fish—thousands of fish. Soon they will come within reach of the nets. The north wind's blowing, and the mullet are running before it out of the sound, running down along the coast, as mullet have done for thousands upon thousands of years.

Half a dozen gulls are mewing above the water. That means the mullet are coming. The gulls don't want the mullet; they want the minnows that are milling about in alarm as the larger fish move through the shallows. The mullet are coming down just outside the breakers, traveling about as fast as a man could walk on the beach. The lookout has marked the school. He walks toward the boat, keeping opposite the fish, signaling their course to the crew by waving his arms.

The men brace their feet against the thwarts of the boat and strain to the oars, pulling the boat in a wide semicircle to the shore. The net of heavy twine spills silently and steadily into the water over the stern and cork floats bob in the water in the wake of the boat. Ropes from one end of the net are held by half a dozen men on shore.

There are mullet in the water all around the boat. They cut the surface with their back fins; they leap and fall. The men lean harder to the oars, pulling for the shore to close the net before the school can escape. Once in the last line of surf and in water not more than waist-deep, the men jump into the water. The boat is seized by willing hands and is dragged out on the beach.

The shallow water in which the mullet are swimming is a pale, translucent green, murky with the loose sand which the waves are stirring up. The mullet are excited by their return to the sea with its bitter salt waters. Under the powerful drive of instinct they move together in the first lap of a journey that will take them far from the coastal shallows, into the blue haze of the sea's beginnings.

A shadow looms in the green, sun-filled water in the path of the mullet. From a dim, gray curtain the shadow resolves itself into a web of slender, crisscross bars. The first of the mullet strike the net, back water with their fins, hesitate. Other fish are crowding up from behind, nosing at the net. As the first waves of panic pass from fish to fish they dash shoreward, seeking a way of escape. The ropes held by fishermen on the shore have been drawn in so that the netting wall extends into water too shallow for a fish to swim. They run seaward, but meet the circle of the net that is growing smaller, foot by foot, as the men on shore and in water up to their knees brace themselves in the sliding sand and pull on the ropes—pull against the weight of water—against the strength of the fish.

As the net is closed and gradually drawn in to shore, the press of fish in the seine becomes greater. Milling in frantic efforts to find a way of escape, the mullet drive with all their combined strength of thousands of pounds against the seaward arc of the net. Their weight and the outward thrust of their bodies lift the net clear of the bottom, and the mullet scrape bellies on the sand as they slip under the net and race into deep water. The fishermen, sensitive to every movement of the net, feel the lift and know they are losing fish. They strain the harder, till muscles crack and backs ache. Half a dozen men plunge out into water chin-deep, fighting the surf to tread the lead line and hold the net on the bottom. But the outer circle of cork floats is still half a dozen boat lengths away.

Of a sudden the whole school surges upward. In a turmoil of flying spray and splashing water mullet by the hundred leap over the cork line. They pelt against the fishermen, who turn their backs to the fish raining about them. The men strive desperately to lift the cork line above the water so that the fish will fall back into the circle when they strike the net.

Two piles of slack netting are growing on the beach, the heads of many small fishes no longer than a man's hand caught in the meshes. Now the ropes attached to the lead lines are drawn in faster and the net takes on the shape of a huge, elongated bag, bulging with fish. As the bag is drawn at last into the shallow fringe of the surf the air crackles with a sound like the clapping of hands as a thousand head of mullet, with all the fury of their last strength, flap on the wet sand.

The fishermen work quickly to take the mullet from the net and toss them into the waiting boats. By a dexterous shake of the net, they toss on the beach the small fish that are gilled in the seine. There are young sea trout and pompano, mullet of the last year's spawning, young ceros and sheepshead and sea bass.

Soon the bodies of the young fish—too small to sell, too small to eat—litter the beach above the water line, the life oozing from them for want of means to cross a few yards of dry sand and return to the sea. Some of the small bodies the sea would take away later; others it would lay up carefully beyond reach of the tides among the litter of sticks and seaweeds, of shells and sea-oats stubbles. Thus the sea unfailingly provides for the hunters of the tide lines.

After the fishermen had made two more hauls and then, as the tide neared the full, had gone away with laden boats, a flock of gulls came in from the outer shoals, white against the graying sea, and feasted on the fish. As the gulls bickered among themselves over the food, two smaller birds in sleek, black plumage walked warily among them, dragging fish up on the higher beach to devour them. They were fish crows, who took their living from the edge of the water, where they found dead crabs and shrimps and other sea refuse. After sundown the ghost crabs would come in legions out of their holes to swarm over the tide litter, clearing away the last traces of the fish. Already the sand hoppers had gathered and were busy at their work of reclaiming to life in their own beings the material of the fishes' bodies. For in the sea, nothing is lost. One dies, another lives, as the precious elements of life are passed on and on in endless chains.

All through the night, as the lights in the fishing village went out one by one and fishermen gathered around their stoves because of the chill north wind, mullet were passing unmolested through the inlet and running westward and southward along the coast, through black water on which the wave crests were like giant fish's wakes, silver in the light of the moon.

BOOK 2

THE GULL'S WAY

6
MIGRANTS OF THE SPRING SEA

Between the Chesapeake Capes and the elbow of Cape Cod the place where the continent ends and the true sea begins lies from fifty to one hundred miles from the tide lines. It is not the distance from shore, but the depth, that marks the transition to the true sea; for wherever the gently sloping sea bottom feels the weight of a hundred fathoms of water above it, suddenly it begins to fall away in escarpments and steep palisades, descending abruptly from twilight into darkness.

In the blue haze of the continent's edge the mackerel tribes lie in torpor during the four coldest months of winter, resting from the eight months of strenuous life in the upper waters. On the threshold of the deep sea they live on the fat stored up from a summer's rich feeding, and toward the end of their winter's sleep their bodies begin to grow heavy with spawn.

In the month of April the mackerel are roused from their sleep as they lie at the edge of the continental shelf, off the Capes of Virginia. Perhaps the currents that drift down to bathe the resting places of the mackerel stir in the fish some dim perception of the progress of the ocean's seasons—the old, unchanging cycle of the sea. For weeks now the cold, heavy surface water—the winter water—has been sinking, slipping under and displacing the warmer bottom water. The warm water is rising, carrying into the surface rich loads of phosphates and nitrates from the bottom. Spring sun and fertile water are wakening the dormant plants to a burst of activity, of growth and multiplication. Spring comes to the land with pale, green shoots and swelling buds; it brings to the sea a great increase in the number of simple, one-celled plants of microscopic size, the diatoms. Perhaps the currents bring down to the mackerel some awareness of the flourishing vegetation of the upper waters, of the rich pasturage for hordes of crustaceans that browse in the diatom meadows and in their turn fill the water with clouds of their goblin-headed young. Soon fishes of many kinds will be moving through the spring sea, to feed on the teeming life of the surface and to bring forth their own young.

Perhaps, also, the currents moving over the place where the mackerel lie carry a message of the inpouring of fresh water as ice and snow dissolve in floods to rush down the coastal rivers to the sea, diminishing ever so slightly its bitter saltiness and attracting the spawn-laden fishes by the lesser density. But however the feeling of awakening spring comes to the dormant fishes, the mackerel stir in swift response. Their caravans begin to form and to move through the dim-lit water, and by thousands and hundreds of thousands they set out for the upper sea.

About a hundred miles beyond the place where the mackerel winter, the sea rises out of the deep, dark bed of the open Atlantic and begins its own climb up over the muddy sides of the continental slope. In utter blackness and stillness the sea climbs those hundred miles, rising from depths of a mile or more until black begins to fade to purple, and purple to deep blue, and blue to azure.

At one hundred fathoms the sea rolls over a sharp edge—the rim of the bowl formed by the foundations of the continent—and starts up the gentler acclivity of the continental shelf. Over the shelving edge of the continent, the sea contains for the first time roving herds of fishes that browse over the fertile undersea plains, for in the deep abyss there are only small, lean fishes hunting singly or in small bands for the sparse food. But here the fishes have rich pasturage—meadows of plantlike hydroids and moss animals, clams and cockles that lie passive in the sand; prawns and crabs that start up and dart away before the rooting snout of a fish, like a rabbit before a hound.

Now small, gasoline-engined fishing boats move over the sea and here and there the water pours through the meshes of miles of gill-net webbing suspended from floats or resists the drag of otter trawls over the sandy floor beneath. And now for the first time the gulls' white wings are patterned in numbers on the sky above, for the gulls—except the kittiwakes—hug the fringes of the sea, feeling uneasy on the open ocean.

As the sea comes in over the continental shelf it meets a series of shoals that run parallel to the coast. In the fifty to one hundred miles to tidewater the sea must hurdle each of these shoals or chains of shoals, climbing up the sides of the hills from the surrounding valleys to shelly plateaus a mile or so wide, then on the shoreward side descending again into the deeper shadows of another valley. The plateaus are more fertile than the valleys in the thousand-odd kinds of back-boneless animals that fishes live on, and so more and larger fish herds browse on them. Often the water above the shoals is especially rich in the moving clouds of small plants and animals of many different kinds that drift with the currents or swim feebly about in search of food—the wanderers or plankton of the sea.

The mackerel do not follow the road over the hills and valleys of the sea's floor as they leave their wintering grounds and turn shoreward. Instead, as though in eagerness to reach at once the sun-lit upper water, they climb steeply the hundred-fathom ascent to the surface. After four months in the gloom of deep water the mackerel move in excitement through the bright waters of the surface layers. They thrust their snouts out of the water as they swim and behold once more the gray expanse of sea cupped in the paleness of arching sky.

Where the mackerel come to the surface there is no sign by which to distinguish the great sea out of which the sun rises from the lesser sea into which it sets; but without hesitation the schools turn from the deep-blue saline water of the open sea and move toward the coastal waters, paled to greenness by the fresh inpouring of the rivers and bays. The place they seek is a great, irregular patch of water that runs from south by west to north by east, from the Chesapeake Capes to southward of Nantucket. In some places it is only twenty miles from shore, in others fifty or more—the spawning grounds in which, from ancient times, the Atlantic mackerel have shed their eggs.

Throughout all the latter part of April mackerel are rising from off the Virginia Capes and hurrying shoreward. There is a stir of excitement in the sea as the spring migration begins. Some of the schools are small; some are as much as a mile wide and several miles long. By day the sea birds watch them rolling landward like dark clouds across the green of the sea; but at night they pour through the water like molten metal, as by their movements they disturb the myriad luminescent animals of the plankton.

The mackerel are voiceless and they make no sound; yet their passage creates a heavy disturbance in the water, so that schools of launce and anchovies must feel the vibrations of an approaching school a long way off and hurry in apprehension through the green distances of the sea; and it may be that the stir of their passage is felt on the shoals below—by the prawns and crabs that pick their way among the corals, by the starfish creeping over the rocks, by the sly hermit crabs, and by the pale flowers of the sea anemone.

As the mackerel hurry shoreward they swim in tier above tier. Throughout those weeks when the fish are rolling in from the open sea the scattered shoals between the edge of the continent and the shore are often darkened as the earth was once dimmed by the passing of another living cloud—the flights of the passenger pigeons.

In time the shoreward-running mackerel reach the inshore waters, where they ease their bodies of their burden of eggs and milt. They leave in their wake a cloud of transparent spheres of infinitesimal size, a vast, sprawling river of life, the sea's counterpart of the river of stars that flows through the sky as the Milky Way. There are known to be hundreds of millions of eggs to the square mile, billions in an area a fishing vessel could cruise over in an hour, hundreds of trillions in the whole spawning area.

After spawning, the mackerel turn toward the rich feeding grounds that lie to seaward of New England. Now the fish are bent only on reaching the waters they knew of old, where the small crustaceans called Calanus move in red clouds through the water. The sea will care for their young, as it cares for the young of all other fishes, and of oysters and crabs and starfish, of worms and jellyfish and barnacles.

7
BIRTH OF A MACKEREL

So it came about that Scomber, the mackerel, was born in the surface waters of the open sea, seventy miles south by east from the western tip of Long Island. He came into being as a tiny globule no larger than a poppy seed, drifting in the surface layers of pale-green water. The globule carried an amber droplet of oil that served to keep it afloat and it carried also a gray particle of living matter so small that it could have been picked up on the point of a needle. In time this particle was to become Scomber, the mackerel, a powerful fish, streamlined after the manner of his kind, and a rover of the seas.

The parents of Scomber were fish of the last big wave of mackerel migration that came in from the edge of the continental shelf in May, heavy with spawn and driving rapidly shoreward. On the fourth evening of their journey, in a flooding current straining to landward, the eggs and milt had begun to flow from their bodies into the sea. Somewhere among the forty or fifty thousand eggs that were shed by one of the female fish was the egg that was to become Scomber.

There could be scarcely a stranger place in the world in which to begin life than this universe of sky and water, peopled by strange creatures and governed by wind and sun and ocean currents. It was a place of silence, except when the wind went whispering or blustering over the vast sheet of water, or when sea gulls came down the wind with their high, wild mewing, or when whales broke the surface, expelled the long-held breath, and rolled again into the sea.

The mackerel schools hurried on into the north and east, their journey scarcely interrupted by the act of spawning. As the sea birds were finding their resting places for the night on the dark water plains, swarms of small and curiously formed animals stole into the surface waters from hills and valleys lying in darkness far below. The night sea belonged to the plankton, to the diminutive worms and the baby crabs, the glassy, big-eyed shrimp, the young barnacles and mussels, the throbbing bells of the jellyfish, and all the other small fry of the sea that shun the light.

It was indeed a strange world in which to set adrift anything so fragile as a mackerel egg. It was filled with small hunters, each of which must live at the expense of its neighbors, plant and animal. The eggs of the mackerel were jostled by the newly hatched young of earlier spawning fishes and of shellfish, crustaceans, and worms. The larvae, some of them only a few hours old, were swimming alone in the sea, busily seeking their food. Some snatched out of the water with pincered claws anything small enough to be overpowered and swallowed; others seized any prey less swift and agile than themselves in biting jaws or sucked into cilium-studded mouths the drifting green or golden cells of the diatoms.

The sea was filled, too, with larger hunters than the microscopic larvae. Within an hour after the parent mackerel had gone away, a horde of comb jellies rose to the surface of the sea. The comb jellies, or ctenophores, looked like large gooseberries, and they swam by the beating of plates of fused hairs or cilia, set in eight bands down the sides of the transparent bodies. Their substance was scarcely more than that of sea water, yet each of them ate many times its own bulk of solid food in a day. Now they were rising slowly toward the surface, where the millions of new-spawned mackerel eggs drifted free in the upper layers of the sea. They twirled slowly back and forth on the long axes of their bodies as they came, flashing a cold, phosphorescent fire. Throughout the night the ctenophores flicked the waters with their deadly tentacles, each a slim, elastic thread twenty times the length of the body when extended. And as they turned and twirled and flashed frosty green lights in the black water, jostling one another in their greed, the drifting mackerel eggs were swept up in the silken meshes of the tentacles and carried by swift contraction to the waiting mouths.

Often during this first night of Scomber's existence the cold, smooth body of a ctenophore collided with him or a searching tentacle missed by a fraction of an inch the floating sphere in which the speck of protoplasm had already divided into eight parts, thus beginning the development by which a single fertile cell would swiftly be transformed into an embryo fish.

Of the millions of mackerel eggs drifting alongside the one that was to produce Scomber, thousands went no farther than the first stages of the journey into life until they were seized and eaten by the comb jellies, to be speedily converted into the watery tissue of their foe and in this reincarnation to roam the sea, preying on their own kind.

Throughout the night, while the sea lay under a windless sky, the decimation of the mackerel eggs continued. Shortly before dawn the water began to stir to a breeze from the east and in an hour was rolling heavily under a wind that blew steadily to the south and west. At the first ruffling of the surface calm the comb jellies began to sink into deep water. Even in these simple creatures, which consist of little more than two layers of cells, one inside the other, there exists the counterpart of an instinct of self-preservation, causing them in some way to sense the threat of destruction which rough water holds for so fragile a body.

In the first night of their existence more than ten out of every hundred mackerel eggs either had been eaten by the comb jellies or, from some inherent weakness, had died after the first few divisions of the cell.

Now, the rising up of a strong wind blowing to southward brought fresh dangers to the mackerel eggs, left for the time being with few enemies in the surface waters about them. The upper layers of the sea streamed in the direction urged upon them by the wind. The drifting spheres moved south and west with the current, for the eggs of all sea creatures are carried helplessly wherever the sea takes them. It happened that the southwest drift of the water was carrying the mackerel eggs away from the normal nursery grounds of their kind into waters where food for young fish was scarce and hungry predators abundant. As a result of this mischance fewer than one egg in every thousand was to complete its development.

On the second day, as the cells within the golden globules of the eggs multiplied by countless divisions, and the shieldlike forms of embryo fish began to take shape above the yolk spheres, hordes of a new enemy came roving through the drifting plankton. The glassworms were transparent and slender creatures that cleaved the water like arrows, darting in all directions to seize fish eggs, copepods, and even others of their own kind. With their fierce heads and toothed jaws they were terrible as dragons to the smaller beings of the plankton, although as men measure they were less than a quarter of an inch long.

The floating mackerel eggs were scattered and buffeted by the dartings and rushes of the glassworms, and when the driftings of current and tide carried them away to other waters a heavy toll of the mackerel had been taken as food.

Again the egg that contained the embryonic Scomber had drifted unscathed while all about him other eggs had been seized and eaten. Under the warm May sun the new young cells of the egg were stirred to furious activity—growing, dividing, differentiating into cell layers and tissues and organs. After two nights and two days of life, the threadlike body of a fish was taking form within the egg, curled halfway around the globe of yolk that gave it food. Already a thin ridge down the mid-line showed where a stiffening rod of cartilage—forerunner of a backbone—was forming; a large bulge at the forward end showed the place of the head, and on it two smaller outpushings marked the future eyes of Scomber. On the third day a dozen V-shaped plates of muscle were marked out on either side of the backbone; the lobes of the brain showed through the still-transparent tissues of the head; the ear sacs appeared; the eyes neared completion and showed dark through the egg wall, peering sightlessly into the surrounding world of the sea. As the sky lightened preparatory to the fifth rising of the sun a thin-walled sac beneath the head—crimson tinted from the fluid it contained—quivered, throbbed, and began the steady pulsation that would continue as long as there was life within the body of Scomber.

Throughout that day development proceeded at a furious pace, as though in haste to make ready for the hatching that was soon to come. On the lengthening tail a thin flange of tissue appeared—the fin ridge from which a series of tail finlets, like a row of flags stiff in the wind, was later to be formed. The sides of an open groove that traversed the belly of the little fish, beneath and protected by the plate of more than seventy muscle segments, grew steadily downward and in midafternoon closed to form the alimentary canal. Above the pulsating heart the mouth cavity deepened, but it was still far short of reaching the canal.

Throughout all this time the surface currents of the sea were pouring steadily to the southwest, driven by the wind and carrying with them the clouds of plankton. During the six days since the spawning of the mackerel the toll of the ocean's predators had continued without abatement, so that already more than half of the eggs had been eaten or had died in development.

It was the nights that had seen the greatest destruction. They had been dark nights with the sea lying calm under a wide sky. On those nights the little stars of the plankton had rivaled in number and brilliance the constellations of the sky. From underlying depths the hordes of comb jellies and glassworms, copepods and shrimps, medusae of jellyfish, and translucent winged snails had risen into the upper layers to glitter in the dark water.

When the first dilution of blackness came in the east, warning of the dawn into which the revolving earth was carrying them, strange processions began to hurry down through the water as the animals of the plankton fled from the sun that had not yet risen. Only a few of these small creatures could endure the surface waters by day except when clouds deflected the fierce lances of the sun.

In time Scomber and the other baby mackerel would join the hurrying caravans that moved down into deep green water by day and pressed upward again as the earth swung once more into darkness. Now, while still confined within the egg, the embryonic mackerel had no power of independent motion, for the eggs remained in water of a density equal to their own and were carried horizontally in their own stratum of the sea.

On the sixth day the currents took the mackerel eggs over a large shoal thickly populated with crabs. It was the spawning season of the crabs—the time when the eggs, that had been carried throughout the winter by the females, burst their shells and released the small, goblin-like larvae. Without delay the crab larvae set out for the upper waters, where through successive moltings of their infant shells and transformations of appearance they would take on the form of their race. Only after a period of life in the plankton would they be admitted to the colony of crabs that lived on that pleasant undersea plateau.

Now they hastened upward, each newborn crab swimming steadily with its wandlike appendages, each ready to discern with large black eyes and to seize with sharp-beaked mouth such food as the sea might offer. For the rest of that day the crab larvae were carried along with the mackerel eggs, on which they fed heavily. In the evening the struggle of two currents—the tidal current and the wind-driven current—carried many of the crab larvae to landward while the mackerel eggs continued to the south.

There were many signs in the sea of the approach to more southern latitudes. The night before the appearance of the crab larvae the sea had been set aglitter over an area of many miles with the intense green lights of the southern comb jelly Mnemiopsis, whose ciliated combs gleam with the colors of the rainbow by day and sparkle like emeralds in the night sea. And now for the first time there throbbed in the warm surface waters the pale southern form of the jellyfish Cyanea, trailing its several hundred tentacles through the water for fish or whatever else it might entangle. For hours at a time the ocean seethed with great shoals of salpae—thimble-sized, transparent barrels hooped in strands of muscle.

On the sixth night after the spawning of the mackerel the tough little skins of the eggs began to burst. One by one the tiny fishlets, so small that the combined length of twenty of them, head to tail, would have been scarcely an inch, slipped out of the confining spheres and knew for the first time the touch of the sea. Among these hatching fish was Scomber.

He was obviously an unfinished little fish. It seemed almost that he had burst prematurely from the egg, so unready was he to care for himself. The gill slits were marked out but were not cut through to the throat, so were useless for breathing. His mouth was only a blind sac. Fortunately for the newly hatched fishlet, a supply of food remained in the yolk sac still attached to him, and on this he would live until his mouth was open and functioning. Because of the bulky sac, however, the baby mackerel drifted upside down in the water, helpless to control his movements.

The next three days of life brought startling transformations. As the processes of development forged onward, the mouth and gill structures were completed and the finlets sprouting from back and sides and underparts grew and found strength and certainty of movement. The eyes became deep blue with pigment, and now it may be that they sent to the tiny brain the first messages of things seen. Steadily the yolk mass shrank, and with its loss Scomber found it possible to right himself and by undulation of the still-rotund body and movement of the fins to swim through the water.

Of the steady drift, the southward pouring of the water day after day, he was unconscious, but the feeble strength of his fins was no match for the currents. He floated where the sea carried him, now a rightful member of the drifting community of the plankton.

8

HUNTERS OF THE PLANKTON

The spring sea was filled with hurrying fishes. Scup were migrating northward from their wintering grounds off the Capes of Virginia, bound for the coastal waters of southern New England where they would spawn. Shoals of young herring moved swiftly just under the surface, rippling the water no more than the passing of a breeze, and schools of menhaden, moving in closely packed formation with bodies flashing bronze and silver in the sun, appeared to the watching sea birds like dark clouds ruffling to a deep blue the smooth sheet of the sea. Mingled with the wandering menhaden and herring were late-running shad, following in along the sea lanes that led to the rivers of their birth, and across the silvery warp of this living fabric the last of the mackerel wove threads of flashing blue and green.

Now, above the water where these hurrying fishes jostled the new-hatched mackerel, there fluttered for the first time that season the little flocks of Oceanites, the petrel, come back to the sea from the far south. The birds moved lightly from place to place on the level plains or the gentle hills of the sea, settling down daintily over some surface-drifting bit of plankton, hovering like butterflies come to sip the nectar from a flower. The little petrels know nothing of the northern winter, for then, in the southern summer, they have gone home from the sea to the far South Atlantic and Antarctic islands where they rear their young.

Sometimes for hours on end the surface of the sea was white with spurting spray as the last of the spring flights of gannets, bound for the rocky ledges of the Gulf of St. Lawrence, plunged from high in the air, pursuing their fish prey far beneath the surface with strong strokes of wings and webbed feet. As the southward drift of the water continued, the gray forms of sharks appeared more often in pursuit of the menhaden schools; the backs of porpoises flashed in the sun; and old, barnacled sea turtles swam at the surface.

As yet Scomber knew little of the world in which he lived. His first food had been the minute, one-celled plants in the water which he drew into his mouth and strained through his gill rakers. Later he had learned to seize the flea-sized crustaceans of the plankton and to dart into their drifting clouds, snapping up the new food with quick snatches. Along with the other young mackerel he spent most of his days many fathoms below the surface and at night rose again to move through dark water that sparkled with the phosphorescent plankton. These movements were made involuntarily as the young fish followed his food, for as yet Scomber knew little of the difference between day and night or the sea's surface and its depths. But sometimes when he climbed with his fins he came into water that was a shining golden green, where moving forms burst upon his vision with swift and terrible vividness.

In the surface waters Scomber first knew the fear of the hunted. On the tenth morning of his life he had lingered in the upper fathoms of water instead of following down into the soft gloom below. Out of the clear green water a dozen gleaming silver fishes suddenly loomed up. They were anchovies, small and herringlike. The foremost anchovy caught sight of Scomber. Swerving from his path, he came whirling through the yard of water that separated them, open-mouthed, ready to seize the small mackerel. Scomber veered away in sudden alarm, but his powers of motion were new-found and he rolled clumsily in the water. In a fraction of a second he would have been seized and eaten, but a second anchovy, darting in from the opposite side, collided with the first and in the confusion Scomber dashed beneath them.

Now he found himself in the midst of the main school of several thousand anchovies. Their silver scales flashed on all sides of him. They bumped and jostled him as he sought in vain to escape. The shoal surged over and beneath and around Scomber, driving furiously onward just under the shining ceiling of the sea. None of the anchovies was now aware of the little mackerel, for the shoal itself was in full flight. A pack of young bluefish had picked up the scent of the anchovies and swung into swift pursuit. In a twinkling they were upon their prey, fierce and ravening as a pack of wolves. The leader of the bluefish lunged. With a snap of razor-toothed jaws he seized two of the anchovies. Two clean-severed heads and two tails floated away. The taste of blood was in the water. As though maddened by it, the bluefish slashed to right and left. They drove through the center of the anchovy school, scattering the ranks of the smaller fish so that they darted in panic and confusion in every direction. Many dashed to the surface and leaped through into the strange element beyond. There they were seized by the hovering gulls, companion fishers of the bluefish.

As the carnage spread, the clear green of the water was slowly clouded with a spreading stain. A strange new taste came with the rusty color and was drawn in by Scomber with the water he passed through mouth and gills. The taste was disquieting to a small fish that had never tasted blood or experienced the lust of the hunter.

When at last pursued and pursuers had passed, the thudding vibrations of the last carnage-mad bluefish were stilled and Scomber's sense cells received once more only the messages of the strong, steady rhythms of the sea. The little mackerel's senses were numbed by the encounter with the swirling, slashing, buffeting monsters. It was in the bright waters of the surface that he had looked upon the racing apparitions, and now that they had passed he made his way down from brightness into green dusk, led down fathom by fathom by the reassuring quality of the gloom which concealed whatever terrors might lurk nearby.

With the descent Scomber came into a cloud of feed, the transparent, big-headed larvae of a crustacean that had spawned in these waters the week before. The larvae moved jerkily through the water, waving the plumelike legs that sprang in two rows from the slender bodies. Scores of young mackerel were feeding on the crustaceans and Scomber joined them. He seized one of the larvae and crushed its transparent body against the roof of his mouth before he swallowed it. Excited and eager for more of the new food he darted among the drifting larvae; and now the sense of hunger possessed him, and the fear of the great fishes was as though it had never been.

As Scomber pursued the larvae in emerald haze five fathoms under the surface he saw a bright flash sweep in a blinding arc across his sphere of vision. Almost instantly the flash was followed by a second blaze of iridescent glitter that curved sharply upward and seemed to thicken as it moved toward a shimmering oval globe above. Once more the thread of the tentacle crept down, all its cilia ablaze in the sunlight. Scomber's instincts warned him of danger, although never before in his larval life had he encountered one of the race of Pleurobrachia, the comb jelly, the foe of all young fishes.

Of a sudden, like a rope swiftly uncoiling from a hand above, one of the tentacles was dropped more than two feet below the inch-long body of the ctenophore, and thus swiftly extended it looped around the tail of Scomber. The tentacle was set with a lateral row of hairlike threads, as barbs grow from the shaft of a bird's feather, but the threads were filmy and tenuous as the strands of a spider's web. All the lateral hairs of the tentacle poured out a gluelike secretion, causing Scomber to become hopelessly entangled in the many threads. He strove to escape, beating the water with his fins and flexing his body violently. The tentacle, contracting and enlarging steadily from the thickness of a hair to that of a thread and then to that of a fishing line, drew him closer and closer to the mouth of the comb jelly. Now he was within an inch of the cold, smooth-surfaced blob of jelly that spun gently in the water. The creature, like a gooseberry in shape, lay in the water with the mouth uppermost, keeping its position by an easy, monotonous beating of the eight rows of ciliated plates or combs. The sun that found its way down from above set the cilia aglow with a radiance that half blinded Scomber as he was drawn up along the slippery body of his foe.

In another instant he would have been seized by the lobelike lips of the creature's mouth and passed into the central sac of its body, there to be digested; but for the moment he was saved by the fact that the ctenophore had caught him while it was still in the midst of digesting another meal. From its mouth there protruded the tail and hind third of a young herring it had caught half an hour before. The comb jelly was greatly distended, for the herring was much too large to be swallowed whole. Although it had tried by violent contractions to force all of the herring past its lips it was unable to do so and had perforce to wait until enough of the fish was digested to make room for the tail. Scomber was held in further reserve, to be eaten after the herring.

In spite of his spasmodic struggles, Scomber was unable to break away from the entangling net of the tentacle hairs, and moment by moment his efforts grew feebler. Steadily and inexorably the contortions of the comb jelly's body were drawing the herring farther into the deadly sac, where digestive ferments worked with marvelous speed to convert the fish tissues, by subtle alchemy, into food for the ctenophore.

Now a dark shadow came between Scomber and the sun. A great, torpedo-shaped body loomed in the water and a cavernous mouth opened and engulfed the ctenophore, the herring, and the entrapped mackerel. A two-year-old sea trout mouthed the watery body of the comb jelly, crushed it experimentally against the roof of its mouth, and spat it out in disgust. With it went Scomber, half dead with pain and exhaustion, but freed from the grip of the dead ctenophore.

When a mass of seaweed that had been torn by the tides from some underlying bed or distant shore floated into Scomber's field of vision, he crept among the fronds and drifted with the weed for a day and a night.

That night as the schools of young mackerel swam near the surface they passed over a sea of death, for ten fathoms below them lay millions of the comb jellies in layer after layer, their bodies almost touching one another, twirling, quivering, tentacles extended and sweeping the water as far as they could reach, sweeping the water clean of every small living thing. Those few young mackerel that strayed down in the night to the level of this solid floor of comb jellies never returned, and when the paling of the water to grayness sent clouds of plankton and many young fishes hurrying down from above, they quickly met their death.

The hordes of the ctenophore Pleurobrachia extended for miles, but fortunately they lay at a deep level and few rose into the upper waters, for in this fashion the sea's creatures are often assorted in layers, one above another. But on the second night the large, lobed ctenophore Mnemiopsis roved through the upper fathoms, and wherever their green lights gleamed in the darkness some small unfortunate of the sea was in peril of its life.

Late that night came the legions of Beroë, the cannibal ctenophore, a sac of pinkish jelly large as a man's fist. The tribe of Beroë was moving out into the coastal waters on a tide of less saline water from a great bay. The sea brought them to the place where the hordes of Pleurobrachia lay twirling and quivering. The big ctenophores fell upon the small ones; they ate them by hundreds and thousands. The loose sacs of their bodies were capable of enormous distension, and scarcely had they been filled when the rapid process of digestion made space for more.

When morning came once more over the sea, the tribes of Pleurobrachia were reduced to a scattered remnant of their former numbers, but a strange stillness lay over the sea where they had been, for in these waters scarcely any living thing remained.

9
THE HARBOR

As the sun entered the sign of the crab, Scomber arrived in the mackerel waters of New England, and with the first spring tides of the month of July he was carried into a small harbor protected from the sea by an outthrust arm of land. From many miles to the southward where the winds and the currents had carried him as a helpless larva, he had returned to the rightful home of young mackerel.

In his third month of life Scomber was more than three inches long. On the journey up the coast the heavy, unmolded lines of a larval fish had been sculptured to a torpedo-shaped body with a hint of power in the shoulders and of speed in the tapering flanks. Now he had put on the sea coat of the adult mackerel. He was clothed in scales, but they were so fine and small that he was soft as velvet to the touch. His back was a deep blue green—the color of the deep places of the sea that Scomber had not yet seen—and over the blue-green background irregular inky stripes ran from the back fin halfway down his flanks. His underparts gleamed of silver, and when the sun found him as he moved just under the surface of the sea he glittered with the colors of the rainbow.

Many young fishes lived in the harbor—cod and herring, mackerel and pollock, cunner and silverside—for the water was rich in food. Twice in every twenty-four hours the flood tide surged in from the open sea through the narrow entrance, flanked on one side by a long sea wall and on the other by a rocky point. The tides came in swiftly, with the push of a great weight of water being forced through a narrow passage, and as they swirled through the cove they carried a wealth of plankton animals mingled with the other small creatures that had been swept off the bottom or plucked from the rocks with the passage of the tide. Twice in every twenty-four hours when the clean, sharp salt water came into the harbor the young fish moved out in excitement to seize the food which the sea had brought them by way of the tide.

Among the young fish in the harbor were several thousand mackerel, who had spent their first weeks of life in many different parts of the coastal waters, but had been brought at last into the harbor by the interplay of currents and by their own wanderings. With the instinct of gregariousness already strong within them, the young mackerel quickly became one school. After the long migrations which each of them had made they were content to live day after day in the waters of the harbor, to rove up and down along the weed-grown sea wall, to feel the spreading of the water over the warm shallows of the cove, and to move out to meet the incoming flood, eager for the swarming copepods and small shrimps it never failed to bring.

The sea coming in through the narrow inlet sucked and swirled over holes scoured out of the bottom and raced in whirlpools and eddies and broke over the rocks in white rips. The tides here moved violently but uncertainly, for the time when the flood turned to the ebb or the ebb to the flood was different within and without the harbor, and what with the push and pull and shifting weight of tides from two sides the water in the inlet race was never still. The rocks of the inlet were matted with creatures that love the swiftly moving current and the ceaseless eddy, and from dark bulges and weed-grown ledges of rock they thrust out eager tentacles and jaws to seize the food animals that swarmed in the water.

Once through the inlet, the sea spread out fanwise into the cove, running swiftly along the old sea wall that formed the eastward rim of the harbor to slap against the wharf pilings and tug at the fishing boats that lay at anchor. Spreading into the western half of the harbor, the water caught the reflection of overhanging scrub oak and cedar and stirred the stones of the shore line to a soft chatter. Toward the northern rim of the cove the water spread out thinly to a sandy beach, wind-rippled above the water line and wave-rippled below.

Over the floor of much of the cove the sea poured through patches of seaweed that grew waist-high to a man. Wherever a rock lay on the bottom one of these underwater gardens grew, and as the floor of the cove was very rocky its pattern as seen from above by the gulls and the terns was mottled darkly with many weed patches. Over the sand-bottomed clearings between the seaweed thickets, the little fishes of the cove poured in restless shoals. The shining green and silver caravans wound in and out, swerving, diverging, and merging again, or at a sudden fright darting away like a shower of silver meteors.

By the same path followed by the sea Scomber came into the harbor, bumped and jostled in the tide rip, whirled and tumbled through the inlet until, seeking quiet water, he found and followed the sandy paths between the rockweed thickets. So he came to the old sea wall, on which the weeds grew in a thick-piled tapestry of browns and reds and greens. As he swam into the swift current that was sweeping the wall a small fish, dark and squat of form, darted out fiercely from the tangle of weeds, causing him to veer away in alarm. The fish was a cunner, like all of its kind a lover of wharfs and harbors. The cunner had lived its whole life in the cove and much of it in the shelter of the sea wall and the fishing wharfs, biting off the barnacles and small mussels that grew on the wharf pilings and finding amphipods and moss animals and scores of other creatures among the seaweeds of pilings and wall. Only the smallest of fish fell prey to the cunner, but by its savage rushes it frightened larger fish away from its feeding places.

Now as Scomber moved up along the wall and came to a dark, quiet place where the deep shadow of a fishing wharf fell across the water, a vast shoal of herring fry burst upon him out of the gloom. The sun struck from their bodies flashes of emerald and silver and bronze. The herring were fleeing from a young pollock that lived in the harbor, terrorizing and preying upon all the smaller fish. As they swirled around Scomber, a new instinct stirred swiftly to life in the young mackerel. He swerved, banked steeply, and seized a young herring athwart its body. His sharp teeth bit deeply into the tender tissues. He carried the herring down into deeper water, just above the swaying ribbons of the weed beds, where he worried it and tore from it several mouthfuls.

As Scomber turned away from his victim, the pollock swung back to look for any herring that might still be lingering in the shadow of the wharf. Seeing Scomber, he swerved down menacingly, but the young mackerel was now too large and swift for him to attack successfully.

The pollock was in his second summer of life, having been born in the winter seas off the Maine coast. As an inch-long fry he had been swept southward in the ocean currents and out to sea, far from his birthplace. Later, as a young fish, pitting the newfound strength of fin and muscle against the sea, he had returned to the coastal shallows, in which he had wandered far to the south of his native waters, preying in season upon the young of other fishes as they schooled close inshore. The pollock was a fierce and ravenous little fish. He could put to rout a school of several thousand cod fry, causing them to scatter in panic and to creep, half paralyzed with fear, into the shelter of seaweeds and rocks.

That morning the pollock killed and ate sixty young herring and in the afternoon, as the launce were coming out of the sand to feed in the flood tide, he played back and forth in the shallows of the cove, slashing at the sharp-nosed, silvery little fish as they emerged. The summer before, when the pollock had been a yearling, the launce had appeared to him the most fearful fish in the sea as they followed and harried the pollock fry, singling out their victims and falling upon them with the ferocity of a pike.

At sunset, Scomber and several score of other small mackerel lay in school formation in blue-gray water a fathom under the surface. For them it was one of the best feeding times of the day, with myriads of the plankton animals streaming by.

The water of the cove lay very still. It was the hour when fishes rise and push their snouts against the surface film, peering out into a strange world of arching sky; when the slow tolling of bells buoyed on distant reefs or shoals comes clear across the water; when the hosts of bottom-living things creep out of burrows and mud tubes and crawl from under stones and loose their grip on wharf pilings to rise into the upper waters.

Before the last shimmer of gold had faded from the surface, Scomber's flanks began to tingle with quick, light vibrations as the water filled with a shoal of clamworms. Nereis, the six-inch clamworm, the bronze water sprite with a scarlet girdle about his middle, rose by the hundred from holes in the sand and under shells of the cove's shallows. By day they lurked in dark recesses under rocks or among the protecting tangles of eel-grass roots, to the end that when a bottom-roving worm or a creeping amphipod moved near they might thrust out their fierce heads armed with amber beaks and seize it. No small bottom dweller could stray near the hole of a nereis and escape death in the waiting jaws.

Although by day the nereis were fierce little beasts of prey in their own domains, with the evening the males among them came forth and swam upward with their fellows to the silver ceiling of the sea. The females remained in their burrows as the night fell fast among the eel-grass roots and the shadows of the overhanging rocks lengthened and grew black. The female nereis wore no scarlet doublets, and the appendages that sprang in a double row from the sides of their bodies were thin and weak, not flattened into swimming paddles as were those of their mates.

A shoal of the big-eyed shrimp had come into the harbor before sunset, followed by more young pollock and, until darkness fell, by a large flock of herring gulls. Although the bodies of the shrimp were transparent they appeared to the gulls like a cloud of moving red dots, for each had a row of brightly colored spots along its sides. Now in the darkness these spots glowed with a strong phosphorescence as the shrimp darted about in the waters of the cove, mingling their fires with the steely green flashes of the ctenophores—creatures that held no further terrors for the young Scomber.

But during that night many strange shapes moved into the water near the fishing wharfs, where the school of young mackerel lay in formation in the black, quiet water. A band of squid, ancestral enemies of all young fishes, had come into the cove. The squid had moved in during the spring from the high seas, their winter home, that they might feed on the hosts of schooling fishes that swarmed over the continental shelf in summer. And as the fishes spawned and their young came to shelter in the

protected harbors, the squid, rapacious in their hunger, pressed in more closely to the land.

Moving against the ebbing tide the squid approached the cove where Scomber and his kind rested. They gave few signs of their coming. They moved more silently than the water that slapped about the wharf pilings. They darted, swift as arrows, through the moving tide, tracing gleaming wakes in the water.

In the chill light of early morning the squid attacked. With the speed of a living bullet the first squid darted into the midst of the school of mackerel, swerved obliquely to the right, and dealt one of the fish an unerring blow just behind the head. The little fish was killed instantly, without ever knowing or having time to fear its foe, for the beak of the squid cut out a clean triangular bite, deep into the spinal cord.

Almost in the same moment half a dozen other squid darted into the mackerel school, but the rush of the first attacker had sent the young fish scattering in all directions. Now the pursuit began, the squid darting among the milling fish, the mackerel dashing and banking and twisting and turning—eluding only by the utmost skill and effort the bottle shapes of the squid that loomed up at terrific speed in the water, tentacles outstretched and grasping.

After the first mad melee Scomber had dashed into the shadow of the wharf and, racing up along the sea wall, had taken shelter under the weeds that grew there. Many other mackerel had done likewise or had darted out into the open water of the cove, scattering widely. Finding that the mackerel had dispersed, the squid dropped to the bottom of the harbor, where their body pigments underwent a subtle change, causing them to match the color of the underlying sand. Soon even the sharpest-eyed fish could not have detected an enemy anywhere about.

The mackerel began to forget their fears and to wander back singly and in little groups to the wharfs where they had been lying, waiting for the turn of the tide. As one by one they swam over the place where a squid lay in motionless invisibility, what had appeared a water-mounded ridge of sand suddenly whirled up from the bottom and seized them.

By these tactics the squid harassed the mackerel all the morning, and only those that remained hidden in the seaweeds of the stone wall were safe from the threat of sudden death.

At the full of the tide the waters of the cove seethed with movement as droves of sand eels or launce raced shoreward. The launce were pursued by a small band of whiting—slender but muscular fish about as long as a man's forearm—with flashing silver underparts and teeth sharp as lancets. The whiting had fallen upon the launce as they emerged from the sand of a shoal two miles to seaward of the cove, to feed on the copepods that the tide was bringing in from farther at sea. The launce fled in terror, not seaward against the tide where they might have found safety by scattering, but with the tide into the cove and into shoaling waters.

As the launce fled, the whiting harried them, driving back and forth across the thousands of slim, finger-long fish. Scomber, lying a foot under water with fins aquiver, felt with suddenly taut nerves the thin staccato vibrations of the racing launce and the heavier roll of the pursuing whiting. The waters about him filled with hurrying shadows. Scomber darted into the shadow of the wharf and hid in the weeds of one of the pilings. Once he would have feared the launce. Now he was as large as they, but the waters were filled with warnings of a hunt and of danger.

As the launce drove deeper into the cove the water began to thin away beneath them, but in their overmastering terror of the whiting they failed to heed the warnings of shoaling water and stranded by hundreds and thousands. The gulls that had followed in expectantly from outside the inlet, sensing what was happening in the seething water below, mewed and squealed and laughed their excitement when they saw the sandy flats beneath them turn to silver. Black-headed laughing gulls and gray-mantled herring gulls came down with flapping wings, plunging shoulder-deep into the water and seizing the launce, screaming threats to the newcomers that dropped down to the feast, although there was an abundance for all.

As the launce piled up inches deep on the shelving beach, the whiting, whirling after in reckless pursuit, drove up on the beach by the dozen, and as the water had now turned to the ebb there was no means of escape. When the tide withdrew the beach was silvered for half a mile with the bodies of the launce, and among them were scattered the larger forms of their pursuers. The squid had followed into the shallow water, attracted by the slaughter, and many of them had stranded while feeding on the hapless launce. Now gulls and fish crows gathered from miles around and, with the crabs and beach fleas, ate of the fish. During that night, wind and tide combined to sweep the beach clean.

The next morning a small bird in bold black and white and ruddy plumage alighted on one of the rocks of the harbor inlet and sat, dozing and dreaming, through fully a quarter of the tide rise before it could rouse itself to pick off and eat some of the small black snails that clung to the rock. The bird was exhausted from fighting the west winds that had threatened to blow it out to sea as it came down the coast from far to the north. It was a ruddy turnstone—one of the first of the great fall flights.

And now as July gave way to August the warm air moving in on the west wind met the cool sea air, and the harbor lay under a dense, dripping fog. From the point a mile down the coast the reedy voice of the foghorn cut through the mist day and night, and bells rang on all the reefs and shoals. For seven days no throb of boat engines came down through the water to the fish in the harbor, for nothing moved over the sea except the gulls, who knew their way in the fog, and the herons, who came to perch on the wharf pilings, guided by the scent of fish in the bait compartments of the boats.

Then the fog passed, and days of blue sky and bluer water followed swiftly one upon another. On these days the flocks of shore birds hurried over the harbor like gusts of autumn leaves, and like windblown leaves their passing betokened the end of summer.

But if knowledge of approaching fall came early to the creatures of shore and marsh, it was slow to awaken in the water world of the cove. When it came it was brought by the southwest wind. Toward the end of August an onshore blow brought rain out of a sky that was grayer than the leaden surface of the harbor. For two days and nights the southwest storm continued, with slanting sheets of water piercing the surface film of the sea with an endless barrage of drops. The rain beat down the incoming and the outgoing tides, so that they rose and lapsed in a waveless surge of water. The flood tides brimmed to the top of the sea wall and swamped many of the fishing boats, so that they wallowed to the bottom, attracting the fishes, who nosed curiously at the strange shapes. All the fish lay deeper under water, and the terns huddled, drenched and disconsolate, on the rocks of the harbor inlet, for with the rain pelting down into the gray opacity of the water they could not see to fish. Unlike the terns, the gulls feasted, for the high storm tides had brought into the harbor much food in the form of injured sea animals and refuse.

After the first day of storm many weeds with narrow, toothed leaves and air vessels like clusters of berries began to appear in the cove, and on the following day the water was filled with floating sargassum weed, which the wind had blown in from the Gulf Stream. Among the fronds of the weed were small and brightly colored fishes that had been carried by the Stream from far to the southward, beginning their long journey as larvae in tropical waters. They had been sheltered by the gulfweed, during the many days and nights of the northward journey, and when the wind blew the weed out of the blue river of warm tropical water the fish accompanied it to the coastal shallows. There most of them would remain, until the coming of unaccustomed cold would abruptly end their lives.

After the storm the waters of the flood tides came in laden with the moon jelly, Aurelia. It was a fateful journey for the beautiful white jellyfish. For a season the ocean had carried them, raised from the algae-grown rocks and shells of the shoreline, where they had begun life as small, plantlike things clinging to stones throughout the winter. In the spring there had budded off from these small creatures a series of flattened discs. These had been quickly transformed into tiny swimming bells, and these to the adult stages. They had lived at the surface when the sun shone and the wind held its breath, often gathering in winding columns miles in length at the meeting places of two currents, where their forms were seen by the gulls, the terns, and the gannets, shimmering in opalescent splendor.

After a time the jellyfish had matured their eggs and then they carried the young in the folds and margins of the tissues that hung like empty sleeves from the underside of the disc. Perhaps the spawning effort had left them weakened, for with bloated tissues and air-inflated egg sacs many of them capsized and floated helplessly in the seas of the late summer. These were set upon by swarms of small crustaceans with hungry jaws and further weakened or destroyed.

Now the southwest storm, kneading the waters deeply, had found the moon jellies. Rough waters seized them and hurried them shoreward. In the jostling and tumbling many tentacles were lost and delicate tissues torn. Every flood tide brought more of the pale discs of the jellyfish into the harbor and cast them up on the rocks of the shoreline. Here their battered bodies became once more a part of the sea, but not until the larvae held within their arms had been liberated into the shallow waters. Thus the cycle came to the full, for even as the substance of the moon jellies was reclaimed for other uses by the sea, the young larvae were settling down for the winter on the stones and shells, so that in the spring a new swarm of tiny bells might rise and float away.

10
SEAWAYS

Now the hours of darkness were as many as the hours of daylight; the sun passed through the constellation of the scales; and September's moon waned to a thin ghost of itself. And as the tides poured through the inlet race into the harbor, creaming with white ripplings over the rocks, and lapsed again to the sea from which they came, they carried away day after day more of the small fish of the harbor. So there came a night when the flood tide stirred in the young mackerel Scomber a strange uneasiness, and on that night the ebb tide, running to the sea, drew him with it. With him went many of the young mackerel who had spent the late summer in the harbor, a school of several hundred cleanly molded young fish each longer than a man's hand. Now they had left behind the pleasant life of the harbor; until death should claim them their world would be open sea.

In the inlet race the mackerel yielded to the eddies and were carried in a swift rush of water past the rocks of the harbor mouth. The water was sharply salt and clean and cold; in its scramble over rocks and shoals it had burst so many rents and tatters in its surface film that it was heavily charged with oxygen. Through this water the mackerel darted in exhilaration, aquiver from their snouts to their last tail finlets—ready and eager for the new life that awaited them. In the inlet the mackerel passed the dark forms of sea bass ranged in the tide, waiting to snap up small crustaceans and sandworms that the water plucked from the rocks or washed out of the holes in the bottom of the channel. The mackerel fled the dark shapes, streaking in swift silver flashes beyond the surging channel where the bass lay, heads into the tide.

Outside the harbor the tide moved with a steadier but heavier pulse, carrying the mackerel out into deeper water. Here the sea came in over shallowing ledges that raised its floor in giant steps from the open basin beyond. Now and again the mackerel felt the drag of current beneath them as they moved above a sandy shoal or weedgrown rocky reef, but ever the undertones of water moving over sand or shells or rock grew more remote as the bottom fell away beneath them, and most of the rhythms and the sound vibrations that came to the hurrying fish were of water and water alone.

The young mackerel moved in a school almost as one fish. None was leader, yet each had a keen awareness of the presence and the movements of all the others, and as those on the margins of the school swung to right or left, or quickened or slackened their pace, so likewise did all the fish of the school.

Now and again the mackerel veered away in sudden alarm from the black shapes of fishing boats that crossed their path, and more than once they darted in momentary panic through the meshes of nets set athwart the tides, being yet too small to become entangled in the twine. Sometimes dark forms lunged at them out of the black water; and once a large squid loomed up and gave them chase, the fish and mollusk darting in and out among a frightened shoal of two-year-old herring, or sperling, on which the squid had been feeding.

Some three miles to seaward of the harbor the mackerel sensed the water shallowing again beneath them as they approached a small island. The island belonged to the sea birds. In season the terns nested on its sands, and the herring gulls brought forth their young under the bushes of beach plum and bayberry and on the flat rocks overlooking the sea. Running out into the sea from the island was a long underwater reef—called by the fishermen The Ripplings—and over it the water was broken into white surges and frothy eddies. As the mackerel passed, scores of pollock were leaping in play in the tide rip, and their bodies gleamed white as the wave froth in the thin light of the risen moon.

When the island and its reef had been left a mile behind, the mackerel school was thrown into sudden panic by the appearance among them of a herd of some half-dozen porpoises which had risen to the surface to blow. The porpoises had been feeding on an underlying sandy shoal, where they were rooting out the launce who had buried themselves there. When the porpoises found themselves among schooling mackerel they slashed at the little fish with their narrow, grinning jaws, killing a few mackerel in sport. But when the school fled in swift alarm through the sea they did not follow, for they had already gorged on the launce to the point of sluggishness.

At early dawn light, the young mackerel, now many miles at sea, came for the first time upon older fish of their own kind. A school of adult mackerel was moving swiftly at the surface of the sea, over which they swept with a heavy rippling. Their snouts were breaking water and their eyes, eager and staring, looked out with water-dimmed sight into the world of air and sky. The two schools—the old fish and the young—merged for a moment of milling confusion as their paths crossed and then continued their separate ways in the sea.

The gulls had come early from their resting places on coastal islands and now they patrolled the sea, their eyes missing nothing that happened in the upper layers and seeing farther down into the water as the sun rose and the shimmer of the level rays faded from the surface. The gulls saw the school of young mackerel swimming a foot under water. Across half a dozen wave hills to the eastward they saw two dark fins, like sickle blades, cutting the water. Because of their elevation the gulls could see that the fins were part of a large fish who drifted just under water, with only the long back fin and the upper blade of the tail fin protruding. The swordfish, who measured eleven feet from the tip of his sword to his tail, often lay idly just beneath the surface, perhaps testing the thrust of the surface ripples with his dorsal fin and so directing his course into the wind. In this way he was certain to meet the shoals of plankton, often accompanied by predatory fish, that drifted with the moving surface water before the wind.

The gulls, who watched the swordfish and the school of young mackerel, now saw a great disturbance approaching from the southeast. An enormous shoal of the big-eyed shrimp was being borne along on the flooding tidal current, which was strengthened by a wind blowing landward. But the shrimp were not browsing on smaller plankton, as the gulls sometimes saw them do, nor drifting peacefully at the surface of the sea. Instead they were fleeing from something that surged through the water with them—open-mouthed and terrible. It was a school of herring, feeding on the shrimp with swift, short rushes. The shrimp were propelling themselves at frantic speed, using all the force of their swimming legs that were flattened into paddlelike blades. And as the space between pursued and pursuers lessened steadily, a shrimp, finding in its transparent body some unused remnant of strength, would fling itself clear of the water just as the jaws of a herring yawned open behind it. But the herring followed relentlessly in pursuit and, though a shrimp might leap half a dozen times into the air, rarely did it escape once a herring had marked it as its victim.

The wind- and current-borne shoal of plankton and the following fish were carried landward; toward them the mackerel swam from the northeast and the swordfish drifted from the northwest. When the fringes of the streaming cloud of plankton reached the mackerel the young fish began to snap eagerly at the shrimp, which were larger food than most of their harbor fare. In a moment, however, they found themselves in the midst of the herring shoal, and the rushes of the larger fishes frightened them and sent them hurrying into deeper water.

The gulls saw the two black fins sink beneath the surface; saw the outlines of the swordfish blur as the large fish dropped deeper into the water and moved beneath the herring. What happened next was partly hidden from the gulls by the seething water and spurting spray; but as they dropped closer and hovered with short wing beats—drawn by awareness of a kill—they could see a great dark shadow that whirled and darted and lunged in a frenzy of attack in the midst of the closely packed ranks of herring. And when the water that foamed to whiteness had grown calmer, more than a score of herring floated at the surface with broken backs and many others swam feebly and listed dizzily, as though they had been injured by glancing blows from the sword. These the great fish now captured easily in its weak-jawed mouth, but many of the dead herring it lost to the gulls, who dropped down to feast on what the swordfish had killed.

When the large fish had killed and eaten to repletion it drifted at the surface of the sea, where the sun-warmed water lulled it to drowsiness. The herring shoals sank into deeper water and the gulls ranged farther to sea, waiting and watching for what might be driven up from below.

Five fathoms down the school of young mackerel had come upon a crimson cloud made up of millions of the small copepod, called Calanus, that were drifting in the tidal current. The mackerel fed on these red crustaceans, which were their favorite food. When the flooding current slackened, hesitated, and grew too weak to carry the plankton with it, the red feed sank into deeper water, followed by the fish. At a depth of only a hundred feet the mackerel came to a gravelly bottom. It was the flat top or plateau of a long undersea hill that curved away to the southward and met another hill coming in from the west, so that the two formed a semicircular ridge with a gully of deep water between. Because of its shape the shoal was known as the Horseshoe to the fishermen, who set their trawl lines over it for haddock, cod, and cusk, and sometimes dragged over it their coneshaped nets or otter trawls.

As the mackerel moved across the shoal they found the bottom beginning to slope away steadily beneath them, and about fifty feet below the highest part of the shoal they came to the edge of the central gully. Three hundred feet below them lay a deep gully floored with soft, sticky mud instead of gravel and broken shell. Many fish called hake lived in the gully, hunting their food in darkness by moving just above the bottom and dragging their long, sensitive fins in the mud. In instinctive fear of deep water, the mackerel school turned and ascended the slope of the shoal. There they moved just above the bottom, in a world that was new and strange to young surface-living fish.

As the mackerel swam over the shoal they were watched by many eyes that looked up from the sand, seeing everything that passed close overhead. They were the eyes of dabs or flounders lying with a thin film of sand over their flat, grayish bodies, so that they were well concealed both from the large predatory fish who would have eaten them and from the shrimps and crabs who scurried over the bottom and were easy prey. The large mouths of the flounders were rimmed with sharp teeth and gaped open as far as the level of the eyes, marking them as occasional fish eaters, but the mackerel were too active and quick-moving to tempt them to rise from their places of concealment and give chase.

Often, as the young mackerel moved over the shoal, a large, heavily built fish with high and pointed back fins would loom up alarmingly close in the water as a haddock swept past and was enveloped again in the gloom. The haddock were very numerous on the Horseshoe, for it was rich in the shelled animals and the spiny-skinned creatures and the tube-dwelling worms that haddock eat. Many times the mackerel came upon small herds of a dozen or more haddock rooting like pigs in the bottom. They were digging out the burrowing worms that had their tunnels deep in the soft sand. As they pushed and dug with their snouts, the black shoulder patches, or "devil's marks," and the black lateral lines stood out vividly in the dim light. The haddock continued their digging, heedless of the young mackerel that darted past with frightened flirts of their tails, for they seldom ate fish when the bottom animals were plentiful.

Once a large, batlike creature fully nine feet across rose from the sand and with a flapping of its thin body passed just above the bottom. So evil and so menacing was its appearance that the school of young mackerel went hurrying upward several fathoms, until the screen of underlying water shut out the sight of the sting ray.

Before a steep ledge of rock they came upon an unfamiliar object dangling in the water. It swayed with the movement of the tide, which ran with great force over the shoal, but it had no motion of its own, although the taste which diffused into the water from it was fishlike. Scomber nosed at the piece of split herring that was bound to a large steel hook and, as he did so, frightened away several small sculpins that had been nibbling at the bait, which was too large for so small a fish to take. Above the hook a thin, dark streak of line stretched away toward a longer line that ran horizontally through the water for a mile over the shoal. As Scomber and his companions ranged over the plateau they saw many of the baited hooks, attached by short lines to the main trawl line. On some of them large fish like haddock were caught, turning and twisting slowly on the hooks that they had swallowed. On one of the hooks was a large cusk, a powerful and heavily built fish some three feet long. The cusk had lived on the shoal, a solitary fish of its kind, spending much of its time hiding among the weeds that grew on the shelving rocks on the outer rim. The scent of the herring bait had drawn it from its hiding place and it had taken the hook. In its struggles the cusk had coiled its powerful body several times about the line.

As the little mackerel fled from the strange sight, the cusk was drawn slowly upward through the water, toward a dim shadow like that of a monster fish on the surface above. The fishermen were running their trawls, rowing from one to another of the lines. If there was a fish on the hook they dispatched it by a blow from a short club, tossing the marketable fish into the bottom of the dory and throwing the other fish back to the sea. It was now an hour after the turn of the tide to the flood, and although the lines had been down only two hours the fishermen had to take them up. On the Horseshoe the currents were very strong, and the line trawls could be set and run only on the slack of the tides.

Now the mackerel came to the seaward rim of the shoal, where the rocky wall fell away in a sheer cliff to the sea bottom some five hundred feet below. All of this outer part of the shoal was solid rock and so it withstood the press of water from the open ocean. Scomber, passing over the rim and above the intense blue water that lay below, found a narrow ledge some twenty feet below the crest of the cliff. Brown, leathery oarweed grew in the crevices and rock layers above the ledge and sent its ribbons streaming out twenty feet or more into the stronger currents that poured by the wall of rock. Scomber nosed his way in among the flat, swaying fronds of the weed and startled a lobster who was resting on the ledge, hidden from the sight of passing fish by the seaweeds. On the underside of her body the lobster carried several thousand eggs attached to the hairs of her swimming legs. The eggs would not hatch until the following spring; meanwhile the lobster was in constant danger of being found by some hungry and inquisitive eel or cunner and stripped of the eggs.

Moving along above the ledge, Scomber suddenly came upon a six-foot rock cod, a two-hundred-pound monster of his kind, who lived on the ledge among the rockweeds. The cod had grown old and very large because of his cunning. He had found the rock ledge above the deep pit of the sea years before and, knowing it instinctively for a good hunting place, he had adopted the ledge for his own, fiercely driving away the other cod. He spent much of his time lying on the ledge, which was in deep purple shadow after the sun had passed the zenith. From this lair he could move out suddenly to seize fishes as they roved along the rock wall. Many fishes met their death in his jaws, among them cunners and hook-eared sculpins, sea ravens with ragged fins, flounders and sea robins, blennies and skates.

Sight of the young mackerel roused the cod from the semi-torpor in which he had lain since the last feeding time and kindled his hunger. He swung his heavy body out from the ledge and climbed steeply to the shoal. Scomber fled before him. As the young mackerel rejoined his fellows who had been lying in an updraught of current from the face of the cliff, the whole school quickened to a sense of alarm and fled away across the shoal as the dark form of the cod loomed into sight at the brink of the rock wall.

The cod roved over the Horseshoe. He fed on all the small creatures—shelled or shell-less—that lived on the bottom or moved above it. He started flounders from where they lay on the sand and sent them darting away before him; he captured small haddock, swinging through the water in swift pursuit of them; he took young fish of his own kind who had recently completed their period of surface life and dropped down to live as true cod on the bottom. He ate dozens of large sea clams, swallowing them whole. After the meats were digested he would expel the shells, although often he carried as many as a dozen of the large shells in his stomach for days, stacked in a neat pile. When

he could find no more sea clams he foraged among the Irish moss that carpeted a flat ledge with a thick, spongy mat, searching for crabs hidden deep within its curling fronds.

A mile away, across the Horseshoe, the mackerel school became aware of a strange disturbance in the water. It was like nothing they had experienced in their life in the harbor, nor during that earlier period, now only the dimmest of memories, when they had drifted with the other plankton at the surface of the sea. It came to them as a heavy, thudding vibration felt with the lateral-line canals along their sensitive flanks. It was not the feel of water vibrations over a rocky reef, nor of waves on a tide rip—yet these sensations were perhaps nearest akin to it of anything the young mackerel had known.

The disturbance grew in strength, and now a group of small cod hurried by, swimming steadily toward the sea rim of the shoal. One by one, and then in groups and small schools, other fish streamed through the water: the great, batlike form of the sting ray, haddock, cod, flounders, a small halibut. All were hurrying toward the edge of the cliff and away from the disturbance that grew until it filled the water with its trembling vibration.

Something vast and dark, like a fish of monstrous and incredible size, its whole forward end a vast, gaping mouth, loomed in the water. At the sight of the cone-shaped net the school of mackerel, which had been confused and irresolute in the presence of the strange vibration and the hurrying fish, suddenly moved as one individual and whirled up and up through water that grew clearer and paler, leaving behind the gloom of the strange world of the shoal, and returning to the surface waters to which they belonged.

As for the fish of the shoal, no such instinct led them up to sun-filled waters and escape. The trawl net had been dragged the length of the Horseshoe and had already scooped up in its cavernous bag thousands of pounds of food fish, as well as quantities of basket starfish, prawns, crabs, clams, cockles, sea cucumbers, and white worm tubes.

The old cod—the cod of the ledge on the cliff side—moved just ahead of the trawl. It was not the first trawl net the monster cod had seen, nor the hundredth. Close behind him the ironbound doors that served to spread the mouth of the net were straining at the long towing cables that stretched away obliquely through the water, stretched up and up toward the vessel steaming a thousand feet in advance of the net.

And now, as the cod swam easily if ponderously above the bottom, he saw that the water before him was changing. It was deepening to the color of water that lies over a great depth. So the cod was accustomed to tell when he was nearing the ledge where he lived above the deep chasm of the sea. The doors of the otter trawl grazed his tail fin. Summoning the great strength dormant in the muscles of his body, he put on a sudden burst of speed, shot out over the blue void, and dropped with precision to his ledge twenty feet below.

Only an instant after the cod passed through the swaying brown thongs of the oarweeds and felt the smooth rock of the ledge beneath his body, the trawl pitched over the edge of the cliff and went tumbling end over end into the deep water below.

11
INDIAN SUMMER OF THE SEA

The spirit of the autumn sea was heard in the voices of the kittiwakes, or frost gulls, who began to arrive in flocks by mid-October. They whirled in thousands over the water, dropping down on arched wings to seize small fish that darted through translucent green. The kittiwakes had come southward from nesting grounds on the cliffs of the Arctic coast and the Greenland ice packs, and with them the first chill breath of winter moved over the graying sea.

There were other signs that autumn had come to the sea. Every day the flights of ocean birds, that in September had poured in thin aerial streams over the coastal waters from Greenland, Labrador, Keewatin, and Baffin Land, swelled in volume as the birds hastened to return to the sea. There were gannets and fulmars, jaegers and skuas, dovekies and phalaropes. Their flocks spread out over all the waters above the continental shelf, where the shoals of surface fishes moved and the plankton herds browsed in the sea.

The gannets were fish eaters that patterned the sky with the white crosses of their bodies as they scanned the sea for prey. Sighting it, they plunged from a hundred feet in the air, and the shock of the heavy body striking the water was broken by a cushion of air sacs under the skin. The fulmars fed on small schooling fish, squid, crustaceans, offal from fishing boats, or any other food that they could seize from the surface, being unable to dive like the gannets. The small dovekies and the phalaropes were eaters of plankton; the jaegers and the skuas lived chiefly on what they could steal from other birds, seldom fishing for themselves.

Few of these birds would see land again until spring. Now they belonged once more to the winter sea, sharing its daylight and darkness, its storms and calms, its sleet and snow and sun and fog.

The yearling mackerel who had left the harbor in late September had at first lived timidly in the open sea, lost in its vastness after the familiar conformations of the harbor. During the three months in the protected cove they had attuned their movements to the rhythms of the tides, feeding on the flood, resting on the ebb. Now the tidal sweep of the surface waters, which here in the open sea, no less than along the coast, yielded to the pull of sun and moon, was almost imperceptible to the young mackerel. For them the tides were lost in the vaster roll of waters. As they roamed the ocean, as yet unfamiliar with its paths of current and varying saltiness, they sought in vain the safe refuges of the harbor, the shadow of the fishing wharfs, the forests of rockweed. Always they must move on into green space.

Scomber and the other yearlings had grown rapidly since they left the harbor, thriving on the rich food of the open ocean. Now in the sixth month of their lives the young fish were from eight to ten inches long—the size fishermen call "tacks." During their first weeks at sea the yearlings moved steadily north and east. In these colder waters the red copepods, their favorite food, tinged miles of ocean with the crimson of their tiny bodies. As the yearlings swung farther offshore and the days of October were marked off by the sun, they found themselves more often among large mackerel, fish of the past dozen years' spawnings. The fall was a time of vast movements of mackerel. The swing of the summer migration, which had carried many of the fish north to the Gulf of St. Lawrence and the coast of Nova Scotia, had passed its climax; flood tide had turned to ebb; and once more the fish moved south.

Slowly the summer warmth was drained from the water. The young crabs, mussels, barnacles, worms, starfish, and crustaceans of scores of species had disappeared from the plankton, for in the ocean spring and summer are the seasons of birth and youth. Only to some of the simplest creatures did the Indian summer of the sea bring a brief and flaring renewal of life, so that they multiplied a millionfold. Among these were the one-celled animals, or protozoa, small as pinpricks, which are among the chief light producers of the sea. Ceratium, the horned one—a blob of protoplasm with three grotesque prongs—sprinkled the night seas of October with silver points of light and so filled the surface waters that over vast areas the sea lay thickened and moved sluggishly under the wind. The little globes of Noctiluca—just visible to the human eye—were each aglitter with submicroscopic grains of light within themselves. During this autumnal period of their great abundance, every fish that moved where the swarms of protozoa were most dense was bathed in light; the waves that broke on reef or shoal spilled liquid fire; and every dip of a fisherman's oar was a flash of a torch in darkness.

On one such night the mackerel came upon an abandoned gill net swaying in the water. The net was buoyed at the surface by floats, and from the cork line it hung down perpendicularly, like a giant tennis net. Its meshes were large enough to allow the yearling mackerel to slip through, although larger fish would have been gilled in the twine. Tonight no fish would have tried to pass through the net, for all its meshes were hung with tiny warning lamps. Luminous protozoa and water fleas and amphipods clung to the wet twine in the dark sea, and the pulse of the ocean stirred from their bodies countless sparks of light. It was as though all the myriad lesser fry of the sea—the plants small as dust motes and the animals tinier than a sand grain—drifting from birth to death in an ocean of infinite size and endless fluidity, seized upon the meshes of the gill net as the one firm reality in their uneasy world and clung to it with protoplasmic hair and cilia, with tentacle and claw. The gill net glowed as though it had life of itself; its radiance shone out into the black sea and down into the darkness below. The light lured many small creatures to rise from deep water and gather on the meshes of the gill net, where they rested all that night in the dark, wide sea.

The mackerel nosed at the net in curiosity and as they bumped the twine all the plankton lamps flared brighter. They followed along its length for more than a mile, for it was set in sections attached one to the other. Other fish were bumping the net. Some were picking off the small sea creatures that clung to it, but none of the fish became gilled.

On moonlight nights the gleam of the moon's radiance would have dimmed the lights of the plankton animals and then many fish, failing to see the net, would have been caught in it. Knowing this, the gill netters fished only in the bright of the moon. This net had been set two weeks before, when the moon was just past the full. For several days two fishermen had tended it from their gasoline-motored launch. Then there had been a night of heavy seas with wind and rain squalls swirling across the water. Since that night the launch had not returned, for it had been wrecked on a shoal about a mile away, and the currents had brought one of its freshly splintered spars and lodged it in the net.

Left to itself, the gill net fished on night after night, and while the moon's light lasted many fish were taken. Dogfish had found them and had torn great holes in the twine as they swarmed in and took the fish. But as the moon's light waned, the plankton lamps burned brighter and no more fish were caught.

Early one morning as the mackerel school swam into the east, Scomber saw above him a long, narrow patch of shadow made by a log that was being carried in the current. He saw the glint of silvery scales from the bodies of several small fish that were moving in the edge of the shadow and swam up to investigate. The log had been part of the cargo of a lumber freighter southward bound from Nova Scotia until it was caught in a northeast gale off the coast of Cape Cod. The freighter, driven onto a shoal, had gone down with all hands, and much of the lumber had been swept ashore in the wind-driven seas. Some, as the storm abated, had been carried offshore and caught in the vast system of ocean currents that swirled clockwise around the fishing banks. The bulk of the drifting log was shelter of the only sort the open sea afforded, and so Scomber joined the little group of fishes, for a period becoming indifferent to the movements of the mackerel school and harking back in his responses to that earlier period of his life when the shadows of fishing wharfs and of anchored boats in the harbor had represented safety from the raids of gulls and squid and large, marauding fishes.

Not long after Scomber had joined the fishes under the log, half a dozen migrating terns settled on it, alighting with a sharp flapping of wings and a scramble of slender toes as they sought a foothold on the surface of the log, already slippery with algae. This was the first time the terns had paused since they had left a beach far to the north the day before. They feared to alight on the water, for although terns take their living from the sea, they are not truly of it. To them the sea was a strange element to which they must often abandon themselves for a brief and frightening instant of contact as they dived for a fish, but not a place on which they would willingly rest their fragile bodies.

The moving wave hills slid under the forward end of the log, tipped it gently skyward, and running swiftly along its length let it slide into the hollows between. As the log lurched and rolled through the sea, seven small fishes followed beneath it, and the terns rode on it like seamen on a raft. As they rested in the midst of the sea, content to let the log carry them out of their course if it would, the terns preened their feathers; they stretched their wings high above their heads, flexing tired muscles; and presently some of them fell asleep.

A little flock of petrels, or Mother Carey's chickens, came down to the water near the log, carrying themselves daintily just above the surface by a pattering of their feet and a fluttering of their wings. Their voices were the thinnest wisps of sound as they whispered over and over their names, *pitterel, pitterel.* The petrels had come down to investigate a dense mass of very small crustaceans who were feeding on the floating body of a dead squid. No sooner had the petrels assembled than a large shearwater came in a great swoop from his patrol in the sky half a mile away and with loud cries plunged in among the small birds. His excited screams brought scores of his kind hurrying to the spot, although a moment before both sky and sea had seemed almost empty of birds. The shearwaters plunged down heavily to the water, striking it with their breasts and flapping their wings. They scattered the petrels as they searched greedily for the food which had attracted the smaller birds to the spot. The first shearwater had already seized the squid, squealing defiance of his companions. Although the squid was too large to swallow whole, the shearwater struggled to gulp it down, for he feared with good reason to relax his grip for an instant.

Suddenly a harsh chattering came down the wind. A dark-brown bird swept through the upper fringes of the cloud of shearwaters. The jaeger whirled past the bird who had possession of the squid, rose into the wind, looped backward, and dropped on the bird. The shearwater plunged and thrashed air and water with his wings, trying to throw off the jaeger and swallow the squid. Suddenly a large piece of the squid fell away and was seized by the jaeger before it could strike the water. After swallowing the prize, the pirate bird sailed off across the water, while all the shearwaters milled about in angry frustration.

By late afternoon a thick mist had closed down over the sea in a blanket spread at about the ordinary cruising height of a shearwater. From golden green the surface waters paled to a gray in which there was neither warmth nor color. The absence of the sun brought the usual rise to the surface of small animals from the lower layers of the sea, and with the lesser fry of the sea came the squid and the fishes that feed on them.

The fog heralded a week of heavy weather, in which the mackerel lived far below the surface, repelled by rough seas. Though swimming deeper than usual, they were still in the upper layers of the sea, for they were passing over a deep basin hollowed out in the continental shelf. Toward the end of the week they approached the outer rim of the basin, where a chain of undersea mountains lay between the coastal waters and the deep Atlantic.

The fall storms had abated, the sun shone again, and the mackerel came up out of the deep gloom to feed once more at the surface. So they passed over a high ridge of the submarine mountain chain. The seas swept over it with a great surge and roll, although they did not break. The movement was unpleasant to the young mackerel, who turned downward to find deeper and quieter water.

A score of the yearling mackerel followed along a dark cliff, where a deep gorge had been hollowed out eons ago. Between the two walls of the undersea valley the sea poured in a green flood. The sun came down through the clear water, leaving the sheer west wall of the cliff in deep-blue shadow. Here and there it lit up a forest of bright-green weed on a shelving ledge and in the dim haze below struck a blaze of color from a spire of jagged rock.

A conger eel lived on one of the ledges of the cliff. The ledge communicated with a deep fissure in the rocks, into which the eel retreated when occasionally it was hard-pressed by some enemy. Sometimes a blue shark, roving through the valley, swerved in to the ledge to attack the thick-bodied conger; or a porpoise came roving along the rock wall, hunting over all the ledges and prying into caves in the cliff for prey. But none of these enemies had been able to capture the conger.

The eel's small eyes saw the mackerel bodies glittering as the little school of fish approached the ledge. The conger gripped the wall of the cave with a muscular tail and drew back its thick body. As the mackerel came abreast of the cave, Scomber swerved out toward the wall of the cliff to investigate a small swarm of amphipods hovering over a fragment of food on a narrow ledge. Instantly the eel loosed its hold on the rock and darted into the open water with a lithe uncoiling of its body. In alarm at the sudden apparition, the mackerel school darted away with a quick acceleration, but Scomber, intent on the amphipods, failed to notice the eel until it was almost upon him.

Down along the cliff raced the two fishes—the mackerel a slim, tapered creature flashing iridescent in the sun; the eel as long as a man is tall and thick and drab as a piece of fire hose. All along the cliff small animals darted back into thickets of weeds or into small holes in the rocks at the passing of the conger, whom they recognized as an enemy. Scomber led the chase up and down along the wall and between spires of projecting rock. At last he dropped down on a weedy ledge. He startled two cunners who had been lying with fins aquiver in a sunlit patch of water just over the margin of the ledge and sent them darting in fear into shelter among the weeds.

Scomber lay very still, his gill covers moving rapidly. Then the currents moving along the rocky wall brought him the taint of conger as the big eel worked its way around the cliff, prying into all the crevices that might shelter a small fish. The scent of his enemy sent Scomber whirling out once more into open water, climbing for the surface. The eel saw the glinting streak of his passage. It turned and gathered speed for the chase, but already it had lost some twenty feet. The conger usually avoided the open water, being a creature of rock ledges and dark, undersea caverns. It hesitated and slackened speed. At this moment its small, deep-set eyes beheld a score of gray fish darting toward it. The eel turned instinctively to race for shelter in its own rock crevice, now left far behind. The school of dogfish bore down upon it. Always ravenous and ever ready to taste blood, the small sharks set upon the eel and in a twinkling had slashed its thick body in a hundred places.

For two days bands of dogfish swarmed in these waters, preying on mackerel, herring, pollock, menhaden, cod, haddock, and every other fish they encountered. On the second day the school to which Scomber belonged, harassed beyond endurance, traveled far to the south and west, above many undersea hills and valleys, and so left behind the shark-infested waters.

That night the mackerel moved through water that was filled with swimming starlets of light. The lights were luminous spots on the bodies of inch-long shrimps, each of which had a pair of light organs under the eyes and twin rows down the sides of their jointed abdomens or tails. When the shrimp flexed their tails in swimming, they could bring the hinder lights to bear on the water beyond and below them and so perhaps were better able to see the small copepods, split-footed shrimps, swimming snails, and one-celled plants and animals which they hunted. Most of the shrimps clutched in their arms, or foremost, bristle-set appendages, a matted pack of the food animals they had already caught by seizing them out of the current set up by the movements of their tails. Following the little darting lights of the shrimps, the mackerel easily found and captured as many of them as they could eat.

At dawn the little sea lamps went out as the first light diluted the water blackness. As they swam toward the sunrise, the mackerel soon found themselves in water that teemed with an enormous shoal of pteropods, or winged snails. As long as the early light lay in level rays across the water, the swarms of pteropods were a hazy, bluish cloud that dimmed the mackerels' vision; but when the sun had been an hour in the sky and its rays came slanting into the sea, the water was filled with a dazzling sparkle and glitter, for the bodies of the pteropods were as transparent and as exquisitely fashioned as the finest glass.

Over miles of sea that morning the mackerel swam through the pteropod shoals, and often they met whales driving open-mouthed through the swarms of mollusks. The mackerel, whom the whales did not seek, fled from the huge, dark forms of the whales; while the winged snails, who were being captured in millions, knew nothing of the monsters who hunted them. Eternally occupied with the quest for food, they browsed peacefully in the sea, unaware of the terrible hunter until the great jaws closed over them and the water rushed away in a torrent through the plates of whalebone.

Swimming down through the school of pteropods, Scomber saw the gleam of a very large fish moving in the water beneath him and felt the heavy roll of displaced water from its wake. But the fish passed from sight as quickly as it had come, and once more Scomber was aware only of feeding mackerel and of the small and glass-clear forms of swimming snails. Then suddenly he felt the great disturbance that troubled the water a few fathoms below him and sensed that mackerel were racing upward from somewhere near the lower fringes of the school. A dozen large tuna had attacked the school of feeding mackerel, having first dropped below the smaller fish to force them to the surface.

As the tuna drove through the milling fish, panic and confusion spread. There was no escape before or behind, nor to right or left. There was none below, where the tuna were. Along with most of his fellows, Scomber climbed up and up. The water was paling as it thinned away above them. Scomber could feel the thudding water vibration of an enormous fish climbing behind him, faster than a small mackerel could climb. He felt the five-hundred-pound tuna graze his flank as it seized the fish swimming beside him. Then he was at the surface, and the tuna were still pursuing. He leaped into the air, fell back, leaped again and again. In the air, birds stabbed at him with their beaks, for the spurting spray was a sign of feeding tuna that brought the gulls hurrying to the spot, to mingle their croaks and screams with the sound of splashing water and of fish bodies falling into the sea.

Now Scomber's leaps were shorter and more labored, and he was falling back with the heaviness of exhaustion. Twice he had barely escaped the jaws of a tuna and many times he had seen one of his companions seized by the attacking fish.

Unseen by mackerel or tuna, a high, black fin was moving over the water from the east. A hundred yards to the southeast of the first fin, two other blades, each as high as a tall man, skimmed rapidly over the sea. Three orcas, or killer whales, were approaching, drawn by the scent of blood.

Then for a space Scomber found the water filled with even more terrifying forms and lashed to a greater confusion as the twenty-foot whales attacked the largest of the tunas, falling upon it like a pack of wolves. Scomber fled from the place where the great fish was plunging and rolling in a vain attempt to escape its enemies. And suddenly he was in water where there were no more tunas to pursue and harry small mackerel, for all of the big fish except the one that was attacked had sped away at sight of the orcas. As he swam down into deeper water, the sea grew calm and still and green again, and now once more he was in the midst of feeding mackerel and saw about him the crystal bodies of swimming snails.

12
SEINE HAUL

That night the sea burned with unusual phosphorescence. Many fish were near the surface, feeding. The chill of November quickened their movements, and as their schools rolled through the water they disturbed the millions of luminous plankton animals, causing them to glow with a fierce luster. So the darkness of the moonless night was broken in many places by flickering patches of light that came and went, flared to brilliance, and died away.

Wandering with half a hundred other yearlings, Scomber saw before him, in darkness pinpricked with silver light, a diffuse glare made by an enormous school of large mackerel, feeding on shrimps, who were pursuing copepods. Thousands of mackerel were drifting slowly with the tide. The whole area covered by the mackerel gleamed mistily, for at every movement of the fish they collided with the myriads of light-producing animals that filled the water.

The yearlings drew closer to the large fish and soon mingled with them. This was a larger school than Scomber had ever known before. All about him were fish—layer upon layer in the water above—layer upon layer below; fish to right and left—fish before him and behind him.

Ordinary the "tacks," or eight- to ten-inch mackerel of the year, would have schooled separately, the division of small fish from large being accomplished by the slower swimming speed of the younger fish. But now that even the larger mackerel—the heavy fish six or eight years old—were moving no faster than the great, sprawling cloud of plankton on which they were preying, the tacks easily kept pace with them, and large and small mackerel schooled together.

The movements of the many fish in the water, the sight of the large mackerel darting, wheeling, turning in darkness, their bodies gleaming with a borrowed light, filled the yearlings with tension and excitement. But so engrossed were the mackerel in feeding that none of them, large or small, was at first aware of the passage through the sea overhead of a luminous streak, like the wake of a giant fish swimming at the surface.

The birds resting on the sea heard the night silence broken by a dull throbbing; some of them that slept more deeply than the others got up from the water only just in time to avoid being struck by the cruising vessel. But neither the startled cry of a fulmar nor the sharp flap of a shearwater's wing could send a message of warning to the fish below.

"Mackerel!" called the lookout at the masthead.

The throb of the engine died away to a scarcely audible heartbeat of sound. A dozen men leaned over the rail of the mackerel seiner, peering into darkness. The seiner carried no light. To do so might frighten the fish. Everywhere was blackness—a thick and velvet blackness in which sky was indistinguishable from water.

But wait! Was there a flicker of light—a pale ghost of flame playing over the water there off the port bow? If there had been such a light it faded away into darkness again and the sea lay in black anonymity—a blank negation of life. But there it came again, and, like a nascent flame in a breeze, or a match cupped in the hands, it kindled to a brilliant glow; it spread into the surrounding darkness; it moved, a gleaming, amorphous cloud, through the water.

"Mackerel," echoed the captain after he had watched the light for several minutes. "Listen!"

At first there was no sound but the soft slap of water against the boat. A sea bird, flying out of darkness into darkness, struck the mast, fell to the deck with a frightened cry, and fluttered off.

Silence again.

Then came a faint but unmistakable patter like a squall of rain on the sea—the sound of mackerel, the sound of a big school of mackerel feeding at the surface.

The captain gave the order to attempt a set. He himself ascended to the masthead to direct the operations. The crew fell into their places: ten into the seine boat attached to a boom on the starboard side of the vessel; two into the dory that was towed behind the seine boat. The throb of the engine swelled. The vessel began to move in a wide circle, swinging around the glowing patch of sea. That was to quiet the fish; to round them up in a smaller circle. Three times the seiner circled the school. The second circle was smaller than the first and the third was smaller than the second. The glow in the water was brighter now and the patch of light more concentrated.

After the third circling of the school, the fisherman in the stern of the seine boat passed to the fisherman in the dory one end of the 1200-foot net that lay piled in the bottom of the seine boat. The seine was dry, having caught no fish that night. The dory cast off and the men at the oars backed water. Again the vessel began to move, towing the seine boat. Now, as the space between the seine boat and dory lengthened, the net slid steadily over the side of the larger boat. A line buoyed by cork floats stretched across the water between them. From the cork line the net hung down in a vertical curtain of webbing a hundred feet deep, held down in the water by leads in the lower border. The line marked out by corks grew from an arc to a semicircle; from a semicircle it swung to the full circle to round up the mackerel in a space four hundred feet across.

The mackerel were nervous and uneasy. Those on the outside of the school were aware of a heavy movement, as of some large sea creature in the water near them. They felt the wash of its passage through the sea—the heavy wake of displaced water. Some of them saw above them a moving, silver shape, long and oval. Beside it moved two smaller forms, one before the other. The shapes might have been those of a she-whale with two calves following at her side. Fearing the strange monsters, the mackerel feeding at the edge of the school turned in toward the center. So, all around the great body of feeding fish, mackerel were wheeling about and plunging in through the school where they could not see the great, luminous shapes and where the wake of the passage of monstrous bodies was lost in the lesser vibrations of thousands of swimming mackerel.

As once more the sea monsters began to circle their prey, only one of the small forms followed the large shape. The other drifted overhead, splashing in the water as with long fins or flippers. Now as the seine boat traced its lesser streak of flame in the water beside the wider gleaming path of the vessel, the netting spilled into the water in its wake. The netting kindled a confused glitter of showering sparks as it slid into the water and hung like a thin, swaying curtain that glimmered palely, for the plankton animals were already gathering on it. The fish were afraid of the netting wall. As the arc enclosed by the twine swung wide and then little by little closed in a great circle, the mackerel at first drew even more compactly together, each part of the school shrinking away from the netting.

Somewhere near the center of the school, Scomber was confusedly aware of the increasing press of fish about him and of the blinding glare of their bodies, clothed in sea light. For him the net did not exist, for he had not seen its plankton-spangled meshes nor brushed its twine with snout or flanks. Uneasiness filled the water and passed with electric swiftness from fish to fish. All about the circle they began to bunt against the net and to veer off and dash back through the school, spreading panic.

One of the fishermen in the seine boat had been only two years at sea. Not long enough to forget, if he ever would, the wonder, the unslakable curiosity he had brought to his job—curiosity about what lay under the surface. He sometimes thought about fish as he looked at them on deck or being iced down in the hold. What had the eyes of the mackerel seen? Things he'd never see; places he'd never go. He seldom put it into words, but it seemed to him incongruous that a creature that had made a go of life in the sea, that had run the gauntlet of all the relentless enemies that he knew roved through that dimness his eyes could not penetrate, should at last come to death on the deck of a mackerel seiner, slimy with fish gurry and slippery with scales. But after all, he was a fisherman and seldom had time to think such thoughts.

Tonight, as he fed the seine into the water and watched the scintillating light as it sank, he thought of the thousands of fish that were milling about down there. He could not see them; even those in the upper water looked only like streaks of light curveting in darkness—fireworks lost in a black, inverted sky, he thought a little dizzily. His mind's eye saw the mackerel running up to the net, bunting it with their snouts, backing off. They would be big mackerel, he thought, for the fiery streaks in the water gave a hint of their size. By the way the phosphorescent light, like a mass of molten metal, was becoming concentrated in the water, he knew that the bumping into the net and the backing off in alarm must be going on all around the circle, for now the ends of the net were closed. The seine boat had overlapped the dory and the two ends of netting had been brought together.

He helped lift the big leaden weight, fit the three-hundred-pound tom over the pursing lines, and start it sliding down the rope to close the open circle in the bottom of the net. The men were beginning to haul in the long purse lines. He thought of the mackerel down there, entrapped only by their own inability to see the way of escape through the bottom of the net. He thought of the tom sliding down, down, into the sea; of the big brass rings that hung from the lead line coming closer together as the purse line was drawn through them; of the dwindling circle at the bottom. But the way of escape must still be open.

The fish were nervous, he could tell. The streaks in the upper water were like hundreds of darting comets. The glow of the whole mass alternately dulled and kindled again to flame. It made him think of the light from steel furnaces in the sky. He seemed to see far down below the surface where the tom was shoving the rings along ahead of it, and the straining ropes were taking up the slack, and the fish were milling in the water—the fish that still had a way of escape. He could imagine that the big mackerel were getting wild. It was too large a school to have set about; but a skipper always hated to split a school. That was almost sure to send them off into deep water. Surely the big fellows would sound yet—would dive down through that shrinking circle straight toward the bottom of the sea, carrying the whole school with them.

He turned away from the water and with his hand felt the pile of wet rope in the bottom of the seine boat, trying to feel—for he could not see—the amount of rope piled up there and trying to guess how much was still to come in before the seine would be pursed.

A shout from the man at his elbow. He turned back to the water. The light within the circle of net was fading, flickering, dying away to an ashy afterglow, to darkness. The fish had sounded.

He leaned over the gunwales, peering down into dark water, watching the glow fade, seeing in imagination what he could not see in fact—the race and rush and downward whirl of thousands of mackerel. He suddenly wished he could be down there, a hundred feet down, on the lead line of the net. What a splendid sight to see those fish streaking by at top speed in a blaze of meteoric flashes! It was only later, when they had finished the long, wet task of repiling the 1200-foot length of seine in the boat, their hour's heavy work wasted, that he realized what it meant that the mackerel had sounded.

After their mad rush through the bottom of the seine, the mackerel scattered widely in the sea, and only when the night was nearly spent did any of the fish that had known the terror of the circling net feed quietly again in schools.

Before dawn, most of the seiners that had fished these waters during the night had vanished in darkness toward the west. One remained, having had bad luck all night, for out of six sets of the seine her crew had lost the fish five times by sounding. The solitary vessel was the only moving thing on the sea that morning when the east turned gray and the black water came ashimmer with silver light. Her crew was hoping for one more set—waiting for the mackerel whom the night's fishing had sent into deep water to show themselves at the surface at daybreak.

Moment by moment the light grew in the east. It picked out the tall mast and the deckhouse of the seiner; it spilled over the gunwales of the following seine boat and lost itself in the pile of netting, black with sea water. It shone on the mounds of the low wave hills and left their valleys in darkness.

Two kittiwakes came flying out of the dimness and perched on the mast, waiting for fish to be caught and sorted.

A quarter of a mile to the southwest, a dark, irregular patch appeared on the water—schooling mackerel, moving slowly into the east.

Quickly the seiner's course was changed to cross in front of the drifting school. With swift maneuvering of the boats, the net was dropped around it. Working with furious haste, the crew sent the tom plunging down the purse line, hauled in the ropes, closed the bottom of the net. Little by little, the men took in the slack of the seine, working the fish into the bunt or central part of the net where the twine was heaviest. Now the vessel came alongside the seine boat and received and made fast the mass of slack netting.

In the water beside the boat lay the bag of the seine, buoyed by corks fastened to the cork line in groups of three or four. In the net were several thousand pounds of mackerel. Most of the fish were large, but among them were a hundred or more tacks or yearlings that had summered in a New England harbor and were only recently of the open sea. One of them was Scomber.

The bailing net, like a ladle of twine on a long wooden handle, was brought into position over the seine, dipped down into the churning mass of fish, raised by pulleys, and emptied out on deck. Several score of lithe and muscular mackerel flapped on the floorboards and sent a rainbow mist of fine scales into the air.

Something was wrong about the fish in the net. Something was wrong about the way they boiled up from below, almost leaping to meet the bailing net. Fish pursed in a seine usually tried to drive the net down, to sink it by sounding. But these fish were terrorized by something in the water—something they feared more than the great boat monster in the water alongside.

There was a heavy disturbance in the water outside the seine. A small triangular fin and the long lobe of a tail cut the surface. Suddenly there were dozens of fins all about the net. A four-foot fish, slim and gray, with a mouth set well back under the tip of his snout, lunged across the cork line and drove his body among the mackerel, slashing and biting.

Now all the dogfish of the pack tore at the seine in ravenous fury, eager to seize the mackerel inside. Their razor-sharp teeth ripped the stout twine as if it had been gauze, and great holes appeared in the net. There was a moment of indescribable confusion, in which the space circumscribed by the cork line became a seething vortex of life—a maelstrom of leaping fish, of biting teeth, of flashing green and silver.

Then, almost as suddenly as it had whirled up, the vortex subsided. In a swift draining away of the turmoil and confusion, the mackerel poured through the holes in the seine, fleet as darting shadows, and lost themselves in the sea.

Among the mackerel who escaped both the seine and the raiding dogfish was the yearling Scomber. By evening of the same day, following older fish and directed by overmastering instinct, he had migrated many miles to seaward of the waters frequented by gill netters and seiners. He was traveling far below the surface, the pale waters of the summer sea forgotten, and was swimming down through deepening green along sea roads new and strange to him. Always he moved south and west. He was going to a place he himself had never known—the deep, quiet waters along the edge of the continental shelf, off the Capes of Virginia.

There, in time, the winter sea received him.

BOOK 3

RIVER AND SEA

13
JOURNEY TO THE SEA

There is a pond that lies under a hill, where the threading roots of many trees—mountain ash, hickory, chestnut oak, and hemlock—hold the rains in a deep sponge of humus. The pond is fed by two streams that carry the runoff of higher ground to the west, coming down over rocky beds grooved in the hill. Cattails, bur reeds, spike rushes, and pickerel weeds stand rooted in the soft mud around its shores and, on the side under the hill, wade out halfway into its waters. Willows grow in the wet ground along the eastern shore of the pond, where the overflow seeps down a grass-lined spillway, seeking its passage to the sea.

The smooth surface of the pond is often ringed by spreading ripples made when shiners, dace, or other minnows push against the tough sheet between air and water, and the film is dimpled, too, by the hurrying feet of small water insects that live among the reeds and rushes. The pond is called Bittern Pond, because never a spring passes without a few of these shy herons nesting in its bordering reeds, and the strange, pumping cries of the birds that stand and sway in the cattails, hidden in the blend of lights and shadows, are thought by some who hear them to be the voice of an unseen spirit of the pond.

From Bittern Pond to the sea is two hundred miles as a fish swims. Thirty miles of the way is by narrow hill streams, seventy miles by a sluggish river crawling over the coastal plain, and a hundred miles through the brackish water of a shallow bay where the sea came in, millions of years ago, and drowned the estuary of a river.

Every spring a number of small creatures come up the grassy spillway and enter Bittern Pond, having made the two-hundred-mile journey from the sea. They are curiously formed, like pieces of slender glass rods shorter than a man's finger. They are young eels, or elvers, that were born in the deep sea. Some of the eels go higher into the hills, but a few remain in the pond, where they live on crayfish and water beetles and catch frogs and small fishes and grow to adulthood.

Now it was autumn and the end of the year. From the moon's quarter to its half, rains had fallen, and all the hill streams ran in flood. The water of the two feeder streams of the pond was deep and swift and jostled the rocks of the streambeds as it hurried to the sea. The pond was deeply stirred by the inrush of water, which swept through its weed forests and swirled through its crayfish holes and crept up six inches on the trunks of its bordering willows.

The wind had sprung up at dusk. At first it had been a gentle breeze, stroking the surface of the pond to velvet smoothness. At midnight it had grown to a half gale that set all the rushes to swaying wildly and rattled the dead seed heads of the weeds and plowed deep furrows in the surface waters of the pond. The wind roared down from the hills, over forests of oak and beech and hickory and pine. It blew toward the east, toward the sea two hundred miles away.

Anguilla, the eel, nosed into the swift water that raced toward the overflow from the pond. With her keen senses she savored the strange tastes and smells in the water. They were the bitter tastes and smells of dead and rain-soaked autumn leaves, the tastes of forest moss and lichen and root-held humus. Such was the water that hurried past the eel, on its way to the sea.

Anguilla had entered Bittern Pond as a finger-long elver ten years before. She had lived in the pond through its summers and autumns and winters and springs, hiding in its weed beds by day and prowling through its waters by night, for like all eels she was a lover of darkness. She knew every crayfish burrow that ran in honeycombing furrows through the mud-bank under the hill. She knew her way among the swaying, rubbery stems of spatterdock, where frogs sat on the thick leaves; and she knew where to find the spring peepers clinging to grass blades, bubbling shrilly, where in spring the pond overflowed its grassy northern shore. She could find the banks where the water rats ran and squeaked in play or tusseled in anger, so that sometimes they fell with a splash into the water—easy prey for a lurking eel. She knew the soft mud beds deep in the bottom of the pond, where in winter she could lie buried, secure against the cold—for like all eels she was a lover of warmth.

Now it was autumn again, and the water was chilling to the cold rains shed off the hard backbones of the hills. A strange restiveness was growing in Anguilla the eel. For the first time in her adult life, the food hunger was forgotten. In its place was a strange, new hunger, formless and ill-defined. Its dimly perceived object was a place of warmth and darkness—darker than the blackest night over Bittern Pond. She had known such a place once—in the dim beginnings of life, before memory began. She could not know that the way to it lay beyond the pond outlet over which she had clambered ten years before. But many times that night, as the wind and the rain tore at the surface film of the pond, Anguilla was drawn irresistibly toward the outlet over which the water was spilling on its journey to the sea. When the cocks were crowing in the farmyard over the hill, saluting the third hour of the new day, Anguilla slipped into the channel spilling down to the stream below and followed the moving water.

Even in flood, the hill stream was shallow, and its voice was the noisy voice of a young stream, full of gurglings and tricklings and the sound of water striking stone and of stone rubbing against stone. Anguilla followed the stream, feeling her way by the changing pressure of the swift water currents. She was a creature of night and darkness, and so the black water path neither confused nor frightened her.

In five miles the stream dropped a hundred feet over a rough and boulder-strewn bed. At the end of the fifth mile it slipped between two hills, following along a deep gap made by another and larger stream years before. The hills were clothed with oak and beech and hickory, and the stream ran under their interlacing branches.

At daybreak Anguilla came to a bright, shallow riffle where the stream chattered sharply over gravel and small rubble. The water moved with a sudden acceleration, draining swiftly toward the brink of a ten-foot fall, where it spilled over a sheer rock face into a basin below. The rush of water carried Anguilla with it, down the steep, thin slant of white water and into the pool. The basin was deep and still and cool, having been rounded out of the rock by centuries of falling water. Dark water mosses grew on its sides and stoneworts were rooted in its silt, thriving on the lime which they took from the stones and incorporated in their round, brittle stems. Anguilla hid among the stoneworts of the pool, seeking a shelter from light and sun, for now the bright shallows of the stream repelled her.

Before she had lain in the pool for an hour another eel came over the falls and sought the darkness of the deep leaf beds. The second eel had come from higher up in the hills, and her body was lacerated in many places from the rocks of the thin upland streams she had descended. The newcomer was a larger and more powerful eel than Anguilla, for she had spent two more years in fresh water before coming to maturity.

Anguilla, who had been the largest eel in Bittern Pond for more than a year, dived down through the stoneworts at sight of the strange eel. Her passage swayed the stiff, limy stems of the chara and disturbed three water boatmen that were clinging to the chara stems, each holding its position by the grip of a jointed leg, set with rows of bristles. The insects were browsing on the film of desmids and diatoms that coated the stems of the stoneworts. The boatmen were clothed in glistening blankets of air which they had carried down with them when they dived through the surface film, and when the passing of the eel dislodged them from their quiet anchorage they rose like air bubbles, for they were lighter than water.

An insect with a body like a fragment of twig supported by six jointed legs was walking over the floating leaves and skating on the surface of the water, on which it moved as on strong silk. Its feet depressed the film into six dimples, but did not break it, so light was its body. The insect's name meant "a marsh treader," for its kind often lived in the deep sphagnum moss of bogs. The marsh treader was foraging, watching for creatures like mosquito larvae or small crustaceans to move up to the surface from the pool below. When one of the water boatmen suddenly broke through the film at the feet of the marsh treader, the twiglike insect speared it with the sharp stilettos projecting beyond its mouth and sucked the little body dry.

When Anguilla felt the strange eel pushing into the thick mat of dead leaves on the floor of the pool, she moved back into the dark recess behind the waterfall. Above her the steep face of the rock was green with the soft fronds of mosses that grew where their leaves escaped the flow of water, yet were always wet with fine spray from the falls. In spring the midges came there to lay their eggs, spinning them in thin, white skeins on the wet rocks. Later when the eggs hatched and the gauzy-winged insects began to emerge from the falls in swarms, they were watched for by bright-eyed little birds who sat on overhanging branches and darted open-mouthed into the clouds of midges. Now the midges were gone, but other small animals lived in the green, water-soaked thickets of the moss. They were the larvae of beetles and soldier flies and crane flies. They were smooth-bodied creatures, lacking the grappling hooks and suckers and the flattened, stream-molded bodies that enabled their relatives to live in the swift currents draining to the brink of the falls overhead or a dozen feet away where the pool spilled its water into the streambed. Although they lived only a few inches from the veil of water that dropped sheer to the pool, they knew nothing of swift water and its dangers; their peaceful world was of water seeping slow through green forests of moss.

The beginning of the great leaf fall had come with the rains of the past fortnight. Throughout the day, from the roof of the forest to its floor, there was a continuous downdrift of leaves. The leaves fell so silently that the rustle of their settling to the ground was no louder than the thin scratching of the feet of mice and moles moving through their passages in the leaf mold.

All day flights of broad-winged hawks passed down along the ridges of the hills, going south. They moved with scarcely a beat of their outspread wings, for they were riding on the updrafts of air made as the west wind struck the hills and leaped upward to pass over them. The hawks were fall migrants from Canada that had followed down along the Appalachians for the sake of the air currents that made the flight easier.

At dusk, as the owls began to hoot in the woods, Anguilla left the pool and traveled downstream alone. Soon the stream flowed through rolling farm country. Twice during the night it dropped over small milldams that were white in the thin moonlight. In the stretch below the second dam, Anguilla lay for a time under an overhanging bank, where the swift currents were undercutting the heavy, grassy turf. The sharp hiss of the water over the slanting boards of the dam had frightened her. As she lay under the bank the eel that had rested with her in the pool of the waterfall came over the milldam and passed on downstream. Anguilla followed, letting the current take her bumping and jolting over the shallow riffles and gliding swiftly through the deeper stretches. Often she was aware of dark forms moving in the water near her. They were other eels, come from many of the upland feeder creeks of the main stream. Like Anguilla, the other long, slender fishes yielded to the hurrying water and let the currents speed their passage. All of the migrants were roe eels, for only the females ascend far into the fresh-water streams, beyond all reminders of the sea.

The eels were almost the only creatures that were moving in the stream that night. Once, in a copse of beech, the stream made a sharp bend and scoured out a deeper bed. As Anguilla swam into this rounded basin, several frogs dived down from the soft mud bank where they had been sitting half out of the water and hid on the bottom close to the bole of a fallen tree. The frogs had been startled by the approach of a furred animal that left prints like those of human feet in the soft mud and whose small black mask and black-ringed tail showed in the faint moonlight. The raccoon lived in a hole high up in one of the beeches nearby and often caught frogs and crayfish in the stream. He was not disconcerted by the series of splashes that greeted his approach, for he knew where the foolish frogs would hide. He walked out on the fallen tree and lay down flat on its trunk. He took a firm grip on its bark with the claws of his hind feet and left forepaw. The right paw he dipped into the water, reaching down as far as he could and exploring with busy, sensitive fingers the leaves and mud under the trunk. The frogs tried to burrow deeper into the litter of leaves and sticks and other stream debris. The patient fingers felt into every hole and crevice, pushed away leaves and probed the mud. Soon the coon felt a small, firm body beneath his fingers—felt the sudden movement as the frog tried to escape. The coon's grip tightened and he drew the frog quickly up onto the log. There he killed it, washed it carefully by dipping it into the stream, and ate it. As he was finishing his meal, three small black masks moved into a patch of moonlight at the edge of the stream. They belonged to the coon's mate and their two cubs, who had come down the tree to prowl for their night's food.

From force of habit, the eel thrust her snout inquisitively into the leaf litter under the log, adding to the terror of the frogs, but she did not molest them as she would have done in the pond, for hunger was forgotten in the stronger instinct that made her a part of the moving stream. When Anguilla slipped into the central current of water that swept past the end of the log, the two young coons and their mother had walked out onto the trunk and four black-masked faces were peering into the water, preparing to fish the pool for frogs.

By morning the stream had broadened and deepened. Now it fell silent and mirrored an open woods of sycamore, oak, and dogwood. Passing through the woods, it carried a freight of brightly colored leaves—bright-red, crackling leaves from the oaks, mottled green and yellow leaves from the sycamores, dull-red leathery leaves from the dogwoods. In the great wind the dogwoods had lost their leaves, but they held their scarlet berries. Yesterday robins had gathered in flocks in the dogwoods, eating the berries; today the robins were gone south and in their place flurries of starlings swept from tree to tree, chattering and rattling and whistling to one another as they stripped the branches of berries. The starlings were in bright new fall plumage, with every breast feather spear-tipped with white.

Anguilla came to a shallow pool formed when an oak had been uprooted in a great autumn storm ten years before and had fallen across the stream. Oak dam and pool were new in the stream since Anguilla had ascended it as an elver in the spring of that year. Now a great mat of weeds, silt, sticks, dead branches, and other debris was packed around the massive trunk, plastering all the crevices, so that the water was backed up into a pool two feet deep. During the period of the full moon the eels lay in the oak-dam pool, fearing to travel in the moon-white water of the stream almost as much as they feared the sunlight.

In the mud of the pool were many burrowing, wormlike larvae—the young of lamprey eels. They were not true eels, but fishlike creatures whose skeleton was gristle instead of bone, with round, tooth-studded mouths that were always open because there were no jaws. Some of the young lampreys had hatched from eggs spawned in the pool as much as four years before and had spent most of their life buried in the mud flats of the shallow stream, blind and toothless. These older larvae, grown nearly twice the length of a man's finger, had this fall been transformed into the adult shape, and for the first time they had eyes to see the water world in which they lived. Now, like the true eels, they felt in the gentle flow of water to the sea something that urged them to follow, to descend to salt water for an interval of sea life. There they would prey semiparasitically on cod, haddock, mackerel, salmon, and many other fishes and in time would return to the river, like their parents, to spawn and die. A few of the young lampreys slipped away over the log dam every day, and on a cloudy night, when rain had fallen and white mist lay in the stream valley, the eels followed.

The next night the eels came to a place where the stream diverged around an island grown thickly with willows. The eels followed the south channel around the island, where there were broad mud flats. The island had been formed over centuries of time as the stream had dropped part of its silt load before it joined the main river. Grass seeds had taken root; seeds of trees had been brought by the water and by birds; willow shoots had sprung from broken twigs and branches carried down in flood waters; an island had been born.

The water of the main river was gray with approaching day when the eels entered it. The river channel was twelve feet deep and its water was turbid because of the inpouring of many tributary streams swollen with autumn rains. The eels did not fear the gloomy channel water by day as they had feared the bright shallows of the hill streams, and so this day they did not rest but pushed on downstream. There were many other eels in the river—migrants from other tributaries. With the increase in their numbers the excitement of the eels grew, and as the days passed they rested less often, pressing on downstream with fevered haste.

As the river widened and deepened, a strange taste came into the water. It was a slightly bitter taste, and at certain hours of the day and night it grew stronger in the water that the eels drew into their mouths and passed over their gills. With the bitter taste came unfamiliar movements of the water—a period of pressure against the downflow of the river currents followed by slow release and then swift acceleration of the current.

Now groups of slender posts stood at intervals in the river, marking out funnel shapes from which straight rows of posts ran slanting toward the shore. Blackened netting, coated with slimy algae, was run from post to post and showed several feet above the water. Gulls were often sitting on the pound nets, waiting for men to come and fish the nets so that they could pick up any fish that might be thrown away or lost. The posts were coated with barnacles and with small oysters, for now there was enough salt in the water for these shellfish to grow.

Sometimes the sandspits of the river were dotted with small shore birds standing at rest or probing at the water's edge for snails, small shrimps, worms, or other food. The shore birds were of the sea's edge, and their presence in numbers hinted at the nearness of the sea.

The strange, bitter taste grew in the water and the pulse of the tides beat stronger. On one of the ebb tides a group of small eels—none more than two feet long—came out of a brackish-water marsh and joined the migrants from the hill streams. They were males, who had never ascended the rivers but had remained within the zone of tides and brackish water.

In all of the migrants striking changes in appearance were taking place. Gradually the river garb of olive brown was changing to a glistening black, with underparts of silver. These were the colors worn only by mature eels about to undertake a far sea journey. Their bodies were firm and rounded with fat—stored energy that would be needed before the journey's end. Already in many of the migrants the snouts were becoming higher and more compressed, as though from some sharpening of the sense of smell. Their eyes were enlarged to twice their normal size, perhaps in preparation for a descent along darkening sea lanes.

Where the river broadened out to its estuary, it flowed past a high clay cliff on its southern bank. Buried in the cliff were thousands of teeth of ancient sharks, vertebrae of whales, and shells of mollusks that had been dead when the first eels had come in from the sea, eons ago. The teeth, bones, and shells were relics of the time when a warm sea had overlain all the coastal plain and the hard remains of its creatures had settled down into its bottom oozes. Buried millions of years in darkness, they were washed out of the clay by every storm to lie exposed, warmed by sunshine and bathed by rain.

The eels spent a week descending the bay, hurrying through water of increasing saltiness. The currents moved with a rhythm that was of neither river nor sea, being governed by eddies at the mouths of the many rivers that emptied into the bay and by holes in the muddy bottom thirty or forty feet beneath. The ebb tides ran stronger than the floods, because the strong outflow of the rivers resisted the press of water from the sea.

At last Anguilla neared the mouth of the bay. With her were thousands of eels, come down, like the water that brought them, from all the hills and uplands of thousands of square miles, from every stream and river that drained away to the sea by the bay. The eels followed a deep channel that hugged the eastern shore of the bay and came to where the land passed into a great salt marsh. Beyond the marsh, and between it and the sea, was a vast shallow arm of the bay, studded with islands of green marsh grass. The eels gathered in the marsh, waiting for the moment when they should pass to the sea.

The next night a strong southeast wind blew in from the sea, and when the tide began to rise the wind was behind the water, pushing it into the bay and out into the marshes. That night the bitterness of brine was tasted by fish, birds, crabs, shellfish, and all the other water creatures of the marsh. The eels lay deep under water, savoring the salt that grew stronger hour by hour as the wind-driven wall of sea water advanced into the bay. The salt was of the sea. The eels were ready for the sea—for the deep sea and all it held for them. Their years of river life were ended.

The wind was stronger than the forces of moon and sun, and, when the tide turned an hour after midnight, the salt water continued to pile up in the marsh, being blown upstream in a deep surface layer while the underlying water ebbed to the sea.

Soon after the tide turn, the seaward movement of the eels began. In the large and strange rhythms of a great water which each had known in the beginning of life, but that each had long since forgotten, the eels at first moved hesitantly in the ebbing tide. The water carried them through an inlet between two islands. It took them under a fleet of oyster boats riding at anchor, waiting for daybreak. When morning came, the eels would be far away. It carried them past leaning spar buoys that marked the inlet channel and past several whistle and bell buoys anchored on shoals of sand or rock. The tide took them close under the lee shore of the larger island, from which a lighthouse flashed a long beam of light toward the sea.

From a sandy spit of the island came the cries of shore birds that were feeding in darkness on the ebb tide. Cry of shore bird and crash of surf were the sounds of the edge of the land—the edge of the sea.

The eels struggled through the line of breakers, where foam seething over black water caught the gleam of the lighthouse beacon and frothed whitely. Once beyond the wind-driven breakers they found the sea gentler, and as they followed out over the shelving sand they sank into deeper water, unrocked by violence of wind and wave.

As long as the tide ebbed, eels were leaving the marshes and running out to sea. Thousands passed the lighthouse that night, on the first lap of a far sea journey—all the silver eels, in fact, that the marsh contained. And as they passed through the surf and out to sea, so also they passed from human sight and almost from human knowledge.

14
WINTER HAVEN

The night of the next full-moon tide, snow came down the bay on a northwest wind. Mile by mile the blanketing whiteness advanced, covering the hills and valleys and marsh flats of the rivers winding toward the sea. Whirling snow clouds swept across the bay, and all through the night the wind screamed over the water, where the flakes were dropping to instant destruction in the blackness of the bay.

The temperature dropped forty degrees in twenty-four hours, and when the tide went out through the mouth of the bay in the morning it left swiftly congealing pools over all the mud flats where it had spread out thinly, and the last of the ebb did not return to the sea.

The cries of the shore birds—twitter of sandpiper and bell note of plover—were silenced, and only the wind's voice was heard, whining over the levels of salt marsh and tide flat. On the last ebb tide the birds had run at the bay's edge, probing the sand; today they were gone before the blizzard.

In the morning, with the snow still whirling out of the sky, a flock of long-tailed ducks, called old squaws, came out of the northwest before the wind. The long-tails were familiars of ice and snow and wintry wind, and they made merry at the blizzard. They cried noisily to one another as they sighted, through the snowflakes, the tall white shaft of the lighthouse that marked the mouth of the bay and saw beyond it a vast gray sheet that was the sea. The old squaws loved the sea. They would live on it throughout the winter, feeding on the shellfish bars of its shallower waters and resting each night on the open ocean, beyond the surf lines. Now they pitched down out of the blizzard—darker flakes among the snow—into the shallows just outside the great salt marsh at the mouth of the bay. Throughout the morning they fed eagerly on the shellfish beds twenty feet below the surface, diving for the small black mussels.

A few of the bay's shore fish still remained in its deeper holes, off the mouths of its lower rivers. They were sea trout, croakers, spots, sea bass, and flukes. These were the fishes that had summered in the bay and spawned, some of them, over its flats or in its river estuaries or its deep holes; the fish that had escaped gilling in the drift nets that came gliding along the bottom on the ebbing tides; the fish that had missed entrapment in the netting mazes that were called pound nets.

Now the bay's waters were in the grip of winter; ice was sealing all its shallows; and its rivers brought down water from the winter hills. So the fishes turned to the sea, remembering with their whole bodies the gently sloping plain that rolled away from the mouth of the bay; remembering the place of warmth, and quiet water, and blue twilight that lay at the edge of the plain.

On the first night of the blizzard a school of sea trout had been trapped by the cold far up in the shallow bay that lay to seaward of the marsh. The thin water chilled so quickly that the warmth-loving trout were paralyzed by cold and lay on the bottom, half dead. When the tide ebbed to the sea, they were unable to follow, but remained in the thinning water. The next morning ice had formed over the head of the shallow bay and the trout perished by the hundred.

Another school of trout that had lain in deeper water off the salt marsh escaped death by the cold. Two spring tides before, these trout had come down from their feeding grounds higher in the bay and had lain just inside the channel to the sea. There the strong ebb tides brought them the feeling of icy water come down the rivers and drawn off the shallows and mud flats.

The trout moved into a deeper channel that was one of a chain of three valleys shaped like the imprint of a monstrous gull's foot deep in the soft sand of the bay mouth. The floor of the channel led them down, fathom by fathom, into quieter and warmer water, over dense beds of weed that swayed to the tide movements. Here the press of the tides was less than over the slopes of the shoals, with the strongest movement of the flood tides confined to the upper layers of water. The ebbs were the scouring tides that poured down along the floor of the valleys, stirring up the sand and carrying empty cockleshells bumping and rolling down the gentle slopes into the deep valleys.

As the sea trout entered the channel, blue crabs from the upper bay passed beneath them, sidling down the slopes from the shallows, seeking the deep, warm holes to spend the winter. The crabs crept into the thick carpet of seaweeds that grew on the channel floor and sheltered other crabs, shrimps, and small fishes.

The trout entered the channel just before nightfall, at the beginning of the ebb. During the early hours of the night other fish moved into the tide flow through the channel and pressed toward the sea. They swam close to the bottom, advancing through the thickets of weed which swayed to the passage of the myriad fish bodies. The fish were croakers that were coming down from all the surrounding shoals, driven by the cold. They lay in tiers, three or four fish deep, beneath the trout, enjoying the channel water, which was many degrees warmer than the water over the shoals.

In the morning, the light in the channel was like a dense green mist, murky with sand and silt. Ten fathoms overhead the last of the flood tide was pushing to westward the red cone of the nun buoy that marked the beginning of the channel as boats came in from the sea. The buoy strained at its anchor chain and tipped and rolled to the surge of water. The trout had come to the junction of the three channels—the heel or spur of the gull's foot that pointed to the sea.

On the next ebb tide the croakers went out through the channel to the sea, seeking waters that were warmer than the bay. The sea trout lingered.

Near the last of the ebb a flurry of young shad passed through the channel, hurrying seaward. They were finger-long fish with scales like white gold. They were among the last of their kind to leave the bay, in the tributaries of which they had hatched from eggs deposited that spring. Thousands of other young of that year had already passed from the shallow, semi-fresh waters of the bay into the vastness of the sea, which was unknown to them and strange. The young shad moved quickly in the briny water of the bay mouth, excited by the strange taste of salt and by the rhythms of the sea.

Snow had ceased to fall, but the wind still blew out of the northwest, piling up the snow into deep drifts and picking up the unpacked surface flakes to whirl them in fantastic wind shapes. The cold was hard and bitter. All the narrower rivers froze from bank to bank, and the oyster boats were locked in their harbors. The bay lay in a hard rim of ice and snow. With every ebb tide, bringing down new water from the rivers, the cold increased in the channel where the sea trout lay.

On the fourth night after the blizzard, the moonglow was strong on the surface of the water. The wind broke the glow into myriad facets of reflected light, and all the ceiling of the bay was aglow with dancing flakes and shaking streamers of light. That night the trout saw hundreds of fish moving into the deep channel above them and passing seaward as dark shadows beneath the silver screen. The fish were other sea trout that had been lying in a ninety-foot hole ten miles up the bay, part of the channel of an ancient river that once had been drowned by the sea to form the bay. The fish that had been lying in the channel like a gull's foot joined the migrants from the deep hole, and together they passed to the sea.

Outside the channel, the trout came to a place of rolling sand hills. The underwater hills were even less stable than the dunes on a windy coast, for they had no roots of sea oats or dune grass to stay them against the thrust of waves that climbed the slope from the deep Atlantic. Some of the hills lay only a few fathoms under water. At every storm they shifted, tons of sand piling up or washing away during a time as short as a single rising of the tide.

After a day of wandering in the sea dunes, the trout rose to a high and tide-swept plateau that marked the seaward end of the sand-hill region. The plateau was half a mile wide and two miles long and overlooked a steeper slope that rolled down steadily into green depths. The shoal itself lay only thirty feet under the surface. Once a strong tide driven in by a southwest wind had shifted the sand and wrecked a fishing schooner bound for port with a ton of fish in its hold. The wreck of the *Mary B.* still lay on the sands, which had sunk away beneath it. Weeds grew from her spars and her masthead, and their long green tapes streamed into the water, pointing landward on the flood tides and seaward on the ebb.

The *Mary B.* lay partly buried in the sand, listing at a forty-five-degree angle to landward. A thick bed of weeds grew under her sheltered or starboard side. The hatch that had covered her fish hold had been carried away in the breaking up of the vessel when she was wrecked, and now the hold was like a dark cave in the sloping floor of the deck—a sea cave for creatures who loved to hide in darkness. The hold was half-full of the crab-cleaned skeletons of the fish that had not washed out of the hold when the vessel sank. The windows of the deckhouse had been smashed by the waves that drove the *Mary B.* aground. Now the windows were used as passageways by all the small fishes that lived about the wreck, nibbling off its encrusting growths. Silvery lookdown fish, spadefish, and triggerfish moved in endless little processions in and out of the windows.

The *Mary B.* was like an oasis of life in miles of sea desert, a place where myriads of the sea's lesser fry—the small, backboneless animals—found a place of attachment; and the small fish foragers found living food encrusting all the planks and spars; and larger predators and prowlers of the sea found a hiding place.

The sea trout drew near to the dark hulk of the wreck as the last green light was fading to gray. They took some of the small fishes and crabs which they found about the vessel, satisfying the hunger born of the long, swift flight from the cold of the bay. Then they settled for the night near the weedy timbers of the *Mary B.*

The trout school lay in the water over the wreck in the lethargy that passed for sleep. They moved their fins gently to keep their position with relation to the wreck and to each other as the water pressed steadily over the shoal, moving up the slope from the sea.

At dusk the winding processions of small fishes that moved in and out of the deckhouse windows and through holes in the rotting planks dispersed and their members found resting places about the wreck. With the twilight which came early through the winter sea, the larger hunters who lived in and about the *Mary B.* stirred swiftly to life.

A long, snakelike arm was thrust out of the dark cavern of the fishhold, gripping the deck with double rows of suction cups. One after another, arms to the number of eight appeared, gripping the deck as a dark form clambered out of the hold. The creature was a large octopus who lived in the fishhold of the *Mary B.* It glided across the deck and slid into the recess above the lower wall of the deckhouse, where it concealed itself to begin the night's hunting. As it lay on the old, weed-grown planks its arms were never still, but reached out busily in all directions, exploring every familiar crack and crevice for unwary prey.

The octopus had not long to wait before a small cunner, intent on the mossy hydroids which it was nibbling off the planks of the vessel, came grazing along the wall of the deckhouse. The cunner, unsuspicious of danger, drew nearer. The octopus waited, its eyes fixed on the moving form, its groping arms stilled. The small fish came to the corner of the deckhouse, jutting out at a forty-five-degree angle to the sea bottom. A long tentacle whipped around the corner and encircled the cunner with its sensitive tip. The cunner struggled with all its strength to escape the clasp of the suckers that adhered to scales, fins, and gill covers, but it was drawn down swiftly to the waiting mouth and torn apart by the cruel beak, shaped like a parrot's.

Many times that night the waiting octopus seized unwary fish or crabs that strayed within reach of its tentacles, or launched itself out into the water to capture a fish passing at a greater distance. Then it moved by a pumping of its flaccid, saclike body, propelling itself by jets of liquid squirted from its siphons. Rarely did the encircling arms and gripping suction cups miss their mark, and gradually the gnawing hunger in the maw of the creature was assuaged.

When the weeds under the prow of the *Mary B.* were swaying confusedly to the turn of the tide, a large lobster emerged from its hiding place in the weed bed and moved off in a general shoreward direction. On land the lobster's unwieldy body would have weighed thirty pounds, but on the sea bottom it was supported by the water so that the creature moved nimbly on the tips of its four pairs of slender walking legs. The lobster carried the large crushing claws, or chelae, extended before its body, ready to seize its prey or attack an enemy.

Moving up along the vessel, the lobster paused to pick off a large starfish that was creeping over the mat of barnacles that covered the stern of the wreck with a white crust. The writhing starfish was conveyed by the pincer claws of the foremost walking legs to the mouth, where other appendages, composed of many joints and moving busily, held the spiny-skinned creature against the grinding jaws.

After eating part of the starfish, the lobster abandoned it to the scavenger crabs and moved on across the sand. Once it paused to dig for clams, turning over the sand busily. All the while its long, sensitive antennae were whipping the water for food scents. Finding no clams, the lobster moved into the shadows for its night's foraging.

Just before dusk, one of the younger sea trout had discovered the third of the large, predatory creatures that lived in the wreck. The third hunter was Lophius, the angler fish, a squat, misshapen creature formed like a bellows, with a wide gash of a mouth set with rows of sharp teeth. A curious wand grew above the mouth, like a supple fishing rod at the end of which dangled a lure, or leaflike flap of flesh. Over most of the angler's body ragged tatters of skin streamed out into the water, giving the fish the appearance of a rock grown with seaweeds. Two thickened, fleshy fins—more like the flippers of a water mammal than the fins of a fish—grew from the sides of its body, and when the angler fish moved on the bottom it drew itself forward by its fins.

Lophius was lying under the prow of the *Mary B.* when the young trout came upon him. The angler fish lay motionless, his two small, evil eyes directed upward from the top of his flat head. He was partly concealed by seaweed and his outline was largely

obliterated by the rags and tatters of loose skin. To all but the most wary of the fish that moved about the wreck Lophius was invisible. Cynoscion, the sea trout, did not notice the angler fish, but saw instead a small and brightly colored object that dangled in the water about a foot and a half above the sand. The object moved; it rose and fell. So small shrimps or worms or other food animals had moved in the trout's experience, and Cynoscion swam down to investigate. When he was twice his own body's length away, a small spadefish whirled in from the open water and nibbled at the lure. Instantly there was a flash of twin rows of sharp, white teeth where a moment before harmless seaweed had swayed to the tides, and the spadefish disappeared into the mouth of the angler.

Cynoscion darted away in momentary panic at the sudden motion and lay under a rotting deck timber, gill covers moving rapidly to his increased inspiration of water. So perfect was the camouflage of the angler that the trout had not seen his outlines;

the only warnings of danger were the flash of teeth and the sudden disappearance of the spadefish. Three times more as he watched the dangling, jerking lure, Cynoscion saw fishes swim up to investigate it. Two were cunners; one was a lookdown fish, high and compressed of body and silvery of color. Each of the three touched the lure and each disappeared into the maw of the angler.

Then twilight passed into darkness, and Cynoscion saw no more as he lay under the rotting deck timbers. But at intervals as the night wore on he felt the sudden movement of a large body in the water beneath him. After about the middle of the night there was no more movement in the weed bed under the prow of the *Mary B.*, for the angler fish had gone out to forage for bigger game than the few small fishes that came to investigate its lure.

A flock of eiders had come down to rest for the night on the water over the shoal. They had alighted first two miles to landward, but the sea ran in broken swells over the rough terrain beneath them and after the tide turn it foamed on the dark water around the ducks. The wind was blowing onshore, and it fought the tide. The ducks were disturbed in their sleep and flew to the outer edge of the shoal, where the water was quieter, and settled down once more on the seaward side of the breakers. The ducks rode low in the water, like laden fishing schooners, Although they slept, some with their heads under the feathers of their shoulders, they often had to paddle with their webbed feet to keep their positions in the swift-running tide.

As the sky began to lighten in the east and the water above the edge of the shoal grew gray instead of black, the forms of the floating ducks looked from below like dark oval shadows encased in a silvery sheen of air imprisoned between their feathers and the surface film. The eiders were watched from below by a pair of small, malignant eyes that belonged to a creature swimming slowly and with awkward motion through the water—a creature like a great, misshapen bellows.

Lophius was well aware that birds were somewhere near, for the scent and taste of duck were strong in the water that passed over the taste buds covering his tongue and the sensitive skin within his mouth. Even before the growing light had brought the surface shadows within his cone-shaped field of vision, he had seen phosphorescent flashes as the feet of the ducks stirred the water. Lophius had seen such flashes before, and often they had meant that birds were resting on the surface. His night's prowling had brought him only a few moderate-sized fishes, which was far from enough to fill a stomach that could hold two dozen large flounders or threescore herring or could pouch a single fish as large as the angler itself.

Lophius moved closer to the surface, climbing with his fins. He swam under an eider that was separated a little from its fellows. The duck was asleep, bill tucked in its feathers, one foot dangling below its body. Before it could waken to knowledge of its danger it was seized in a sharp-toothed mouth with a spread of nearly a foot. In sudden terror the duck beat the water with its wings and paddled with its free foot, seeking to take off from the surface. By a great exertion of strength it began to rise from the water, but the full weight of the angler hung from its body and dragged it back.

The honking of the doomed eider and the thrashing of its wings alarmed its companions, and with a wild churning of the water the remainder of the flock took off in flight, quickly disappearing into the thin mist that lay over the sea. The duck was bleeding spurts of bright-red blood from a severed leg artery. As its life ebbed away in the bright stream, its struggles grew feeble, and the strength of the great fish prevailed. Lophius pulled the duck under, sinking away from the cloud of reddened water just as a shark appeared in the dim light, attracted by the scent of blood. The angler took the duck to the floor of the shoal and swallowed it whole, for his stomach was capable of enormous distension.

Half an hour later Cynoscion, the sea trout, hunting about the wreck for small fishes, saw the angler returning to his hole under the prow of the *Mary B.*, pulling himself over the bottom by his handlike pectoral fins. He saw Lophius creep into the shadow of the vessel and saw the weeds that waved under the prow part to receive him. There the angler would lie in torpor for several days, digesting his meal.

During the day the water chilled by almost imperceptible degrees, and in the afternoon the ebb tide brought a great flood of cold water from the bay. That evening the sea trout, driven by the cold, left the wreck and ran seaward during the entire night, passing down the plain that sloped steadily away beneath them. They moved over smooth, sandy bottoms, sometimes rising to pass over a mound or shoal of broken shell. They hurried on, resting seldom because of the creeping cold. Hour by hour the water above them deepened.

The eels must have passed this way, through the country of underwater sand hills and down the sloping meadowlands and prairies of the sea.

Often during the next few days the trout were overtaken by other schools of fishes when they paused for rest or food and often they met browsing fish herds of many different kinds. The fish had come from all the bays and rivers of many miles of coastline, fleeing the winter cold. Some had come from far to the north, from the coasts of Rhode Island and Connecticut and the shores of Long Island. These were scup, thin-bodied fish with high, arched backs and spiny fins, covered with platelike scales. Every winter the scup came from New England to the waters off the Capes of Virginia and then returned in spring to spawn in the northern waters and be caught in traps and swiftly encircling seines. The farther the sea trout traveled across the continental shelf, the more often they saw the scup herds in the green haze before them, the large bronze fish rising and sinking as they grubbed on the bottom for worms, sand dollars, and crabs and drifted up a fathom or more to munch their food.

And sometimes there were cod schools, come from Nantucket Shoals to winter in the warmer southern waters. Some of the cod would spawn in this place that seemed alien to their kind, leaving their young to the ocean currents, which might never return them to the northern home of the cod.

The cold increased. It was like a wall moving through the sea across the coastal plain. It was nothing that could be seen or touched; yet it was so real a barrier that no fish would have run back through it any more than if it had been solid as stone. In milder winters the fish would have scattered widely over the continental shelf—the croakers well inshore; the flukes or flounders on all the sandy patches; scup in all the sloping valleys, rich in bottom food; and sea bass over every piece of rocky ground. But this year the cold drove them on, mile after mile, to the edge of the continental shelf—to the edge of the deep sea. There in the quiet water, warmed by the Gulf Stream, they found a winter haven.

Even as the fish were running out across the continental shelf from all the bays and rivers, boats were moving south and out to sea. The boats were squat and ungraceful of line and they pitched and rolled in the winter sea. They were trawlers, come from many northern ports to find the fish in their winter refuge.

Only a decade before, the sea trout, the fluke, the scup, and the croakers had been safe from the fishermen's nets once they had left the bays and sounds. Then, one year, boats had come, dragging nets like long bags. The boats had moved down from the north and out from the coast, towing their nets along the bottom. At first they had taken nothing. Mile by mile, they moved farther out, and finally their nets came up filled with food fishes. The wintering grounds of the shore fish—the summer fish of the bays and river estuaries—had been discovered.

From that time on, the trawlers came every season and took millions of pounds of fishes each year. Now they were on their way, coming down from the northern fishing ports. There were haddock trawlers from Boston and flounder draggers from New Bedford; there were redfish boats from Gloucester and cod boats from Portland. Winter fishing in southern waters is easier than winter fishing on the Scotian Banks or the Grand Banks; easier even than on Georges, or Browns, or the Channel.

But this winter was cold; the bays were icebound, and the sea was gale-ridden. The fish were far out; seventy miles out, a hundred miles out. The fish were deep down in warm water, a hundred fathoms down.

The trawls went over the side, from decks that were slippery with freezing spray. The meshes of the trawl nets were stiff with ice, and all the ropes and the cables groaned and creaked with the frost. The trawls went down through the hundred fathoms of water; down from ice and sleet and heaving sea and screaming wind to a place of warmth and quiet, where fish herds browsed in the blue twilight, on the edge of the deep sea.

15
RETURN

The record of the eels' journey to their spawning place is hidden in the deep sea. No one can trace the path of the eels that left the salt marsh at the mouth of the bay on that November night when wind and tide brought them the feeling of warm ocean water—how they passed from the bay to the deep Atlantic basin that lies south of Bermuda and east of Florida half a thousand miles. Nor is there a clearer record of the journey of those other eel hordes that in autumn passed to the sea from almost every river and stream of the whole Atlantic Coast from Greenland to Central America.

No one knows how the eels traveled to their common destination. Probably they shunned the pale-green surface waters, chilled by wintry winds and bright as the hill streams they had feared to descend by day. Perhaps they traveled instead at mid-depths or followed the contours of the gently sloping continental shelf, descending the drowned valleys of their native rivers that had cut channels across the coastal plain in sunshine millions of years ago. But somehow they came to the continent's edge, where the muddy slopes of the sea's wall fell away steeply, and so they passed to the deepest abyss of the Atlantic. There the young were to be born of the darkness of the deep sea and the old eels were to die and become sea again.

In early February billions of specks of protoplasm floated in darkness, suspended far below the surface of the sea. They were the newly hatched larvae—the only testament that remained of the parent eels. The young eels first knew life in the transition zone between the surface sea and the abyss. A thousand feet of water lay above them, straining out the rays of the sun. Only the longest and strongest of the rays filtered down to the level where the eels drifted in the sea—a cold and sterile residue of blue and ultraviolet, shorn of all its warmth of reds and yellows and greens. For a twentieth part of the day the blackness was displaced by a strange light of a vivid and unearthly blue that came stealing down from above. But only the straight, long rays of the sun when it passed the zenith had power to dispel the blackness, and the deep sea's hour of dawn light was merged in its hour of twilight. Quickly the blue light faded away, and the eels lived again in the long night that was only less black than the abyss, where the night had no end.

At first the young eels knew little of the strange world into which they had come, but lived passively in its waters. They sought no food, sustaining their flattened, leaf-shaped bodies on the residue of embryonic tissue, and so they were the foes of none of their neighbors. They drifted without effort, buoyed by their leafy form and by the balance between the density of their own tissues and that of the sea water. Their small bodies were colorless as crystal. Even the blood that ran in its channels, pumped by hearts of infinitesimal size, was unpigmented; only the eyes, small as black pinpricks, showed color. By their transparency the young eels were better fitted to live in this twilight zone of the sea, where safety from hungry foragers was to be found only in blending with the surroundings.

Billions of young eels—billions of pairs of black, pinprick eyes peering into the strange sea world that overlay the abyss. Before the eyes of the eels, clouds of copepods vibrated in their ceaseless dance of life, their crystal bodies catching the light like dust motes when the blue gleam came down from above. Clear bells pulsated in the water, fragile jellyfish adjusted to life where five hundred pounds of water pressed on every square inch of surface. Fleeing before the descending light, shoals of pteropods, or winged snails, swept down from above before the eyes of the watching eels, their forms glistening with reflected light like a rain of strangely shaped hailstones—daggers and spirals and cones of glassy clearness. Shrimps loomed up—pale ghosts in the dim light. Sometimes the shrimps were pursued by pale fishes, round of mouth and flabby of flesh, with rows of light organs set like jewels on their gray flanks. Then the shrimps often expelled jets of luminous fluid that turned to a fiery cloud to blind and confuse their enemies. Most of the fishes seen by the eels wore silver armor, for silver is the prevailing color or badge of those waters that lie at the end of the sun's rays. Such were the small dragonfish, long and slender of form, with fangs glistening in their opened mouths as they roamed through the water in an endless pursuit of prey. Strangest of all were the fishes, half as long as a man's finger and clothed in a leathery skin, that shone with turquoise and amethyst lights and gleamed like quicksilver over their flanks. Their bodies were thin from side to side and tapered to sharp edges. When enemies looked down from above, they saw nothing, for the backs of the hatchetfish were a bluish black that was invisible in the black sea. When sea hunters looked up from below, they were confused and could not distinguish their prey with certainty, for the mirrorlike flanks of the hatchetfish reflected the blueness of the water and their outlines were lost in a shimmer of light.

The young eels lived in one layer or tier of a whole series of horizontal communities that lay one below the other, from the nereis worms that spun their strands of silk from frond to frond of the brown sargassum weed floating on the surface to the sea spiders and prawns that crawled precariously over the deep and yielding oozes of the floor of the abyss.

Above the eels was the sunlight world where plants grew, and small fishes shone green and azure in the sun, and blue and crystal jellyfish moved at the surface.

Then came the twilight zone where fishes were opalescent or silver, and red prawns shed eggs of a bright orange color, and round-mouthed fishes were pale, and the first light organs twinkled in the gloom.

Then came the first black layer, where none wore silvery sheen or opalescent luster, but all were as drab as the water in which they lived, wearing monotones of reds and browns and blacks whereby they might fade into the surrounding obscurity and defer the moment of death in the jaws of an enemy. Here the red prawns shed deep-red eggs, and the round-mouthed fishes were black, and many creatures wore luminous torches or a multitude of small lights arranged in rows or patterns that they might recognize friend or enemy.

Below them lay the abyss, the primeval bed of the sea, the deepest of all the Atlantic. The abyss is a place where change comes slow, where the passing of the years has no meaning, nor the swift succession of the seasons. The sun has no power in those depths, and so their blackness is a blackness without end, or beginning, or degree. No beating of tropical sun on the surface miles above can lessen the bleak iciness of those abyssal waters that varies little through summer or winter, through the years that melt into centuries, and the centuries into ages of geologic time. Along the floor of the ocean basins, the currents are a slow creep of frigid water, deliberate and inexorable as the flow of time itself.

Down beneath mile after mile of water—more than four miles in all—lay the sea bottom, covered with a soft, deep ooze that had been accumulating there through eons upon eons of time. These greatest depths of the Atlantic are carpeted with red clay, a pumicelike deposit hurled out of the earth from time to time by submarine volcanoes. Mingled with the pumice are spherules of iron and nickel that had their origin on some far-off sun and once rushed millions of miles through interstellar space, to perish in the earth's atmosphere and find their grave in the deep sea. Far up on the sides of the great bowl of the Atlantic the bottom oozes are thick with the skeletal remains of minute sea creatures of the surface waters—the shells of starry Foraminifera and the limy remains of algae and corals, the flintlike skeletons of Radiolaria and the frustules of diatoms. But long before such delicate structures reach this deepest bed of the abyss, they are dissolved and made one with the sea. Almost the only organic remains that have not passed into solution before they reach these cold and silent deeps are the ear bones of whales and the teeth of sharks. Here in the red clay, in the darkness and stillness, lies all that remains of ancient races of sharks that lived, perhaps, before there were whales in the sea; before the giant ferns flourished on the earth or ever the coal measures were laid down. All of the living flesh of these sharks was returned to the sea millions of years before, to be used over and over again in the fashioning of other creatures, but here and there a tooth still lies in the red-clay ooze of the deep sea, coated with a deposit of iron from a distant sun.

The abyss south of Bermuda is a meeting place for the eels of the western and eastern Atlantic. There are other great deeps in the ocean between Europe and America—chasms sunk between the mountain ranges of the sea's floor—but only this one is both deep enough and warm enough to provide the conditions which the eels need for the act of spawning. So once a year the mature eels of Europe set out across the ocean on a journey of three to four thousand miles; and once a year the mature eels of eastern America go out as though to meet them. In the westernmost part of the drifting sea of sargassum weed some of them meet and intermingle—those that travel farthest west from Europe and farthest east from America. So in the central part of the vast spawning grounds of the eels, the eggs and young of two species float side by side in the water. They are so alike in appearance that only by counting with infinite care the vertebrae that make up their backbones and the plates of muscle that flank their spines can they be distinguished. Yet some, toward the end of their period of larval life, seek the coast of America and others the coast of Europe, and none ever stray to the wrong continent.

As the months of the year passed, one by one, the young eels grew, lengthening and broadening. As they grew and the tissues of their bodies changed in density, they drifted into light. Upward passage through space in the sea was like passage through time in the Arctic world in spring, with the hours of sunlight increasing day by day. Little by little the blue haze of midday lengthened and the long nights grew shorter. Soon the eels came to the level where the first green rays, filtering down from above, warmed the blue light. So they passed into the zone of vegetation and found their first food.

The plants that received enough energy for their life processes from the sea-strained residue of sunlight were microscopic, floating spheres. On the cells of ancient brown algae the young eels first nourished their glass-clear bodies—plants of a race that had lived for untold millions of years before the first eel, or the first backboned animal of any kind, moved in the earth's seas. Through all the intervening eons of time, while group after group of living things had risen up and died away, these lime-bearing algae had continued to live in the sea, forming their small protective shields of lime that were unchanged in shape and form from those of their earliest ancestors.

Not only the eels browsed on the algae. In this blue-green zone, the sea was clouded with copepods and other plankton foraging on the drifting plants, and dotted with the swarms of shrimplike animals that fed on the copepods, and lit by the twinkling silver flashes of small fishes that pursued the shrimps. The young eels themselves were preyed upon by hungry crustaceans, squid, jellyfish, and biting worms, and by many fishes who roved openmouthed through the water, straining food through mouth and gill raker.

By midsummer the young eels were an inch long. They were the shape of willow leaves—a perfect shape for drifters in the currents. Now they had risen to the surface layers of the sea, where the black dots of their eyes could be seen by enemies in the bright-green water. They felt the lift and roll of waves; they knew the dazzling brightness of the midday sun in the pure waters of the open ocean. Sometimes they moved in the midst of floating forests of sargassum weed, perhaps taking shelter beneath the nests of flying fishes or, in the open spaces, hiding in the shadow of the blue sail or float of a Portuguese man-of-war.

In these surface waters were moving currents, and where the currents flowed the young eels were carried. All alike were swept into the moving vortex of the north Atlantic drift—the young of the eels from Europe and the young of the eels from America. Their caravans flowed through the sea like a great river, fed from the waters south of Bermuda and composed of young eels in numbers beyond enumeration. In at least a part of this living river, the two kinds or species of eels traveled side by side, but now they could be distinguished with ease, for the young of the American eels were nearly twice as large as their companions.

The ocean currents swept in their great circle, moving from south through west and north. Summer drew to its end. All the sea's crops had been sown and harvested, one by one—the spring crop of diatoms, the swarms of plankton animals that grew and multiplied on the abundant plants, the young of myriad fishes that fed on the plankton herds. Now the lull of autumn was upon the sea.

The young eels were far from their first home. Gradually the caravan began to diverge into two columns, one swinging to the west, one to the east. Before this time there must have been some subtle change in the responses of the faster-growing group of eels—something that led them more and more to the west of the broad river of moving surface water. As the time approached for them to lose the leaflike form of the larva and become rounded and sinuous like their parents, the impulse to seek fresher, shallowing waters grew. Now they found the latent power of unused muscles, and against the urging of wind and current they moved shoreward. Under the blind but powerful drive of instinct, every activity of their small and glassy bodies was directed unconsciously toward the attainment of a goal unknown in their own experience—something stamped so deeply upon the memory of their race that each of them turned without hesitation toward the coast from which their parents had come.

A few eastern-Atlantic eels still drifted in the midst of the western-Atlantic larvae, but none among them felt the impulse to leave the deep sea. All their body processes of growth and development were geared to a slower rate. Not for two more years would they be ready for the change to the eel-like form and the transition to fresh water. So they drifted passively in the currents.

To the east, midway across the Atlantic, was another little band of leaflike travelers—eels spawned a year before. Farther to the east, in the latitude of the coastal banks of Europe, was still another host of drifting eel larvae, these yet a year older and grown to their full length. And that very season a fourth group of young eels had reached the end of their stupendous journey and was entering the bays and inlets and ascending the rivers of Europe.

For the American eels the journey was shorter. By midwinter their hordes were moving in across the continental shelf, approaching the coast. Although the sea was chilled by the icy winds that moved over it, and by the remoteness of the sun, the migrating eels remained in the surface waters, no longer needing the tropical warmth of the sea in which they had been born.

As the young moved shoreward, there passed beneath them another host of eels, another generation come to maturity and clothed in the black and silver splendor of eels returning to their first home. They must have passed without recognition—these two generations of eels—one on the threshold of a new life; the other about to lose itself in the darkness of the deep sea.

The water grew shallower beneath them as they neared the shore. The young eels took on their new form, in which they would ascend the rivers. Their leafy bodies became more compact by a shrinkage in length as well as in depth, so that the flattened leaf became a thickened cylinder. The large teeth of larval life were shed, and the heads became more rounded. A scattering of small pigment-carrying cells appeared along the backbone, but for the most part the young eels were still as transparent as glass. In this stage they were called "glass eels," or elvers.

Now they waited in the gray March sea, creatures of the deep sea, ready to invade the land. They waited off the sloughs and bayous and the wild-rice fields of the Gulf Coast, off the South Atlantic inlets, ready to run into the sounds and the green marshes that edged the river estuaries. They waited off the ice-choked northern rivers that came down with a surge and a rush of spring floods and thrust long arms of fresh water into the sea, so that the eels tasted the strange water taste and moved in excitement toward it. By the hundreds of thousands they waited off the mouth of the bay from which, little more than a year before, Anguilla and her companions had set out for the deep sea, blindly obeying a racial purpose which was now fulfilled in the return of the young.

The eels were nearing a point of land marked by the slim white shaft of a lighthouse. The sea ducks could see it—the piebald old-squaw ducks—when they circled high above the sea on their return every afternoon from inshore feeding grounds, coming down at dusk to the dark water with a great rush and a roar of wings. The whistling swans saw it, too, painted by the sunrise on the green sea beneath them as their flocks swept northward in the spring migration. The leader swans blew a triple note at the sight, for the point of land marked the nearness of the first stop on the swans' long flight from the Carolina Sounds to the great barrens of the Arctic.

The tides were running high with the fullness of the moon. On the ebb tides the taste of fresh water came strongly to the fish that lay at sea, off the mouth of the bay, for all the rivers were in flood.

In the moon's light the young eels saw the water fill with many fish, large and full-bellied and silvery of scale. The fish were shad returned from their feeding grounds in the sea, waiting for the ice to come out of the bay that they might ascend its rivers to spawn. Schools of croakers lay on the bottom, and the roll of their drums vibrated in the water. The croakers, with sea trout and spots, had moved in from their offshore wintering place, seeking the feeding grounds of the bay. Other fish came up into the tide flow and lay with heads to the currents, waiting to snap up the small sea animals that the swiftly moving water had dislodged, but these were bass who were of the sea and would not ascend the rivers.

As the moon waned and the surge of the tides grew less, the elvers pressed forward toward the mouth of the bay. Soon a night would come, after most of the snow had melted and run as water to the sea, when the moon's light and the tide's press would be feeble and a warm rain would fall, mist-laden and bittersweet with the scent of opening buds. Then the elvers would pour into the bay and, traveling up its shores, would find its rivers.

Some would linger in the river estuaries, brackish with the taste of the sea. These were the young mate eels, who were repelled by the strangeness of fresh water. But the females would press on, swimming up against the currents of the rivers. They would move swiftly and by night as their mothers had come down the rivers. Their columns, miles in length, would wind up along the shallows of river and stream, each elver pressing close to the tail of the next before it, the whole like a serpent of monstrous length. No hardship and no obstacle would deter them.

They would be preyed upon by hungry fishes—trout, bass, pickerel, and even by older eels; by rats hunting the edge of the water; and by gulls, herons, kingfishers, crows, grebes, and loons. They would swarm up waterfalls and clamber over moss-grown rocks, wet with spray; they would squirm up the spillways of dams. Some would go on for hundreds of miles—creatures of the deep sea spreading over all the land where the sea itself had lain many times before.

And as the eels lay offshore in the March sea, waiting for the time when they should enter the waters of the land, the sea, too, lay restless, awaiting the time when once more it should encroach upon the coastal plain, and creep up the sides of the foothills, and lap at the bases of the mountain ranges. As the waiting of the eels off the mouth of the bay was only an interlude in a long life filled with constant change, so the relation of sea and coast and mountain ranges was that of a moment in geologic time. For once more the mountains would be worn away by the endless erosion of water and carried in silt to the sea, and once more all the coast would be water again, and the places of its cities and towns would belong to the sea.

Glossary

Abyss. The central deeps of the ocean, enclosed by the steep walls of the continental slope. The floor of the abyss is a vast and desolate plain, lying, on the average, about three miles deep, with occasional valleys or canyons dropping off to depths of five or six miles. The bottom is covered with a deep, soft deposit composed of inorganic clays and of the insoluble remains of minute sea creatures. The abyss is wholly without light and is uniformly cold.

Alga (ăl´-gȧ; pl. algae (-jē)). The algae belong to the first of the four major divisions of the plant kingdom and are the simplest and probably the oldest plants. They do not have true roots, stems, or leaves, but usually consist of a simple, leaflike frond. They range in size from microscopic spheres to giant seaweeds several hundred feet long. (See oarweed.)

Amphipod (ăm´-fĭ-pŏd). Belonging to the same large group as crabs, lobsters, and shrimps, the amphipods comprise a large group of crustaceans whose bodies are flattened from side to side and covered with a polished and flexible cuticle that is divided into sections, allowing them to jump or swim with surprising agility. There are about three thousand species of amphipods, most of which live in the sea or about its edge. Perhaps the most familiar of these are the sand fleas. Caprella often attaches itself by the hind legs to a bit of seaweed and extends its body stiffly, so that it may easily be mistaken for a branch of the weed. It is about half an inch long.

Anchovy (ăn-chō´-vĭ). Anchovies are small, silvery fish of herringlike appearance. They usually travel in schools which are the prey of many larger fishes. The common anchovy or whitebait is from two to four inches long.

Angler fish. The angler is notorious as perhaps the ugliest, most repulsive, and most voracious of fishes. Half of the angler is head, and a good portion of the head is mouth, hence one of its local names: "all-mouth." The angler is found on both sides of the Atlantic and may be as much as four feet long.

Anguilla (ăng-gwĭl´-á). The scientific name of the common eel.

Aurelia (ô-rē´-lĭ-á). A flat, saucer-shaped jellyfish of a white or bluish-white color that may be up to a foot in diameter. Its appearance while swimming has suggested the common name "moon jelly." Unlike many other jellyfishes, it has small and inconspicuous tentacles. The moon jelly is found on both Atlantic and Pacific coasts.

Avens, mountain. A dwarf, hardy shrub of the rose family, also called "wild betony," found in Arctic and north temperate regions. The flowers are large and white, and the leaves are said to be one of the chief foods of the ptarmigan in winter.

Barnacle. In spite of the hard shells that enclose it, the barnacle is not related to oysters and clams, as many people suppose, but is a crustacean and so related to crabs, lobsters, and water fleas. The shells remain open while they are covered by water, and the legs, as delicately feathered as an ostrich plume, are thrust out rhythmically to aerate the blood contained in the filaments and to kick small food animals into the mouth. When the tide ebbs, barnacles that grow between the tide lines close their shells with an audible click.

Basket starfish. A species of starfish with intricately branched arms, on the tips of which it walks. It preys on fishes which are so unfortunate as to venture within the brushlike mass of arms, and is found from eastern Long Island northward, in offshore waters.

Beach flea. (See Sand flea.)

Beroë (bĕr´-ō-ē). One of the larger ctenophores (about four inches long), which feeds largely on its own relatives, often swallowing prey as large as itself. These ctenophores are abundant in New England waters in July and August, appearing at the surface during the warmest part of the day, and dropping to greater depths when the water is cold or rough.

Betony. (See Avens.)

Big-eyed shrimp. So called because of the large eyes which are very conspicuous in the nearly transparent bodies of these shrimplike crustaceans. Especially interesting are the phosphorescent spots which vary in number and arrangement with the species. These shrimps occur at the surface in swarms, usually accompanied by schools of fish and sometimes by immense flocks of gulls. They are often to be seen in tide rips.

Blenny. This small fish lives among seaweeds and stones from the tide lines down to depths of thirty to fifty fathoms or sometimes a little deeper. Its body is elongated and somewhat eel-like, with a fin running almost the entire length of the back.

Brant. Shallow coastal bays are ideal feeding grounds for these black and gray geese, who obtain their favorite food—the roots and lower stems of eel grass—by "tipping up" where the water is shallow enough and pulling up the grass. Their migration routes take them from Virginia and North Carolina to Greenland and the extreme northern Arctic Islands, via Cape Cod, the Gulf of St. Lawrence, and Hudson Bay.

Brown algae. Among the brown algae is a group (called "round lime bearers") whose members wear shields of lime united into a remarkable defensive armor. Remains of these shields are found in very ancient geological deposits, at least as remote as Cambrian time. Present-day forms are practically identical in structure to their prehistoric ancestors.

Bryozoa (brī-ō-zō´-ȧ). Marine and fresh-water animals usually of a delicately branched and mosslike form. Early naturalists considered them plants. Some types form limy crusts of lacelike appearance on stones and seaweeds. The group is a very ancient one.

Byssus thread (bĭs´-ŭs). Certain shellfish, such as clams, mussels, and the like, possess (especially during infancy) a gland capable of secreting a fluid that hardens into a tough thread or cord on contact with seawater. This thread, called the byssus, serves to anchor its owner against the pull of surf or tidal flow.

Calanus (căl´-ȧ-nŭs). A small copepod crustacean (about an eighth of an inch long) that is extremely abundant at certain seasons of the year off the New England coast. Its economic importance is considerable, because it is one of the principal foods of the herring and mackerel, also of the Greenland whale. (See copepod and crustacean.)

Ceratium (sē-rā´-shĭ-ŭm). A single-celled creature about 1/100 of an inch in diameter, claimed by botanists as well as zoologists, but usually considered an animal. It is extremely phosphorescent, and during the periods of its greatest abundance the sea blazes with light when the ceratium is disturbed.

Cero (sē´-rō). A large, silvery fish of the mackerel tribe, found chiefly in southern waters. Another common name is "kingfish." It is a strong and active predator, and often is found among schools of menhaden.

Chara (kā´-rȧ). This fresh-water alga forms underwater meadows in ponds or lakes receiving water from lime-containing soils. The plant is characteristically rough and brittle to the touch because of the carbonate of lime deposited in its tissues and on its surface. In some waters it forms large deposits of marl, a crumbling, limy substance used as a fertilizer for soils deficient in lime. The leaflets grow from the central stem in candelabralike clusters, and the fruiting bodies remind one of translucent Japanese lanterns of pinhead size, some orange and some green.

Chela (kē´-lȧ). The large, pincerlike claw of a lobster, the muscles of which are considered the choicest parts of the animal for eating. It is an effective weapon for defense or attack.

Chitin (kī´-tĭn). A horny substance that forms the harder part of the outer covering of insects, lobsters, crabs, and the like.

Chlorophyll (klō´-rō-fíl). The green coloring matter of plants, which plays an essential part in the manufacture of starches and sugars by the leaves.

Cilium (sĭl´-ĭ-ŭm). A minute, hairlike projection from a cell. Usually occurring in numbers wherever present, cilia set up a current by rhythmic lashing movements. Some one-celled animals and plants and some larvae of higher forms move by cilia.

Cockle. A mollusk with a heart-shaped shell usually sculptured into radiating ridges and handsomely marked both inside and out. The cockle is a much more active shellfish than the related clams, and progresses along the bottom by surprising leaps and tumbles. These are effected by thrusting out a muscular "foot," bending it under the shell, and suddenly straightening it.

Conger eel (kŏng´-gẽr). Conger eels are exclusively marine, reach a weight of fifteen pounds or more in American waters and up to one hundred and twenty-five pounds in European, and are exceedingly voracious.

Continental shelf. The gently sloping bottom of the sea from the tide lines down to depths of approximately one hundred fathoms is known as the continental shelf. In places the continental shelf of the United States is about a hundred miles wide; in others, as off the Florida coast, it is only a few miles wide. Many parts of the present shelf were land in comparatively recent geological times. Most marine commercial fisheries are confined to waters over the shelf. The steeper descent from the edge of the shelf to the oceanic abyss is known as the continental slope.

Copepod (kō´-pē-pŏd). A large subclass of crustaceans (q.v.) all less than two-fifths of an inch long and most of them much smaller. Many are free-swimming members of the plankton; some use the bodies of living animals as homes from which they come and go without detriment to their host; others are parasites on the gills, skin, or flesh of fish. They are one of the most important links in the marine food chain, making plant foods available to the many young fishes and other creatures that feed on them. (See Calanus, for example.)

Crab larva. Newly hatched crabs are transparent, big-headed creatures that bear no resemblance to their parents. As they grow they must shed the hard cuticle that covers them with an unyielding armor, and so they pass through a series of molts, each of which brings them a little closer to a crablike physique. Their early life is spent near the surface, swimming about actively and snapping up smaller creatures from the surrounding water.

Crane fly. An adult crane fly is a long-legged, mosquitolike insect often seen about streams at dusk, or flying about lights after dark. Their larvae live in the water or in moist places.

Croaker. An abundant fish of the Atlantic Coast south of New England, which owes its common name to its ability to make a grunting or croaking sound by drumming with a pair of specialized muscles on its air bladder (a balloonlike sac under the backbone). This drumming may be heard a considerable distance under water. Another common name, used especially in the Chesapeake Bay area, is "hardhead."

Crowberry. A low-growing evergreen shrub of Arctic regions from Alaska to Greenland, found also as far south as the northern United States. Its berries are a favorite food of Arctic birds.

Crustacean. Animals that wear a segmented shell and have segmented legs are arthropods; arthropods that live in the water and breathe by gills are crustaceans. Familiar examples are lobsters, barnacles, shrimps, and crabs.

Ctenophore (tĕn´-ō-fōr). A marine animal much like a jellyfish. Most ctenophores are cylindrical or pear-shaped and swim by the beating of hairlike cilia arranged in eight longitudinal bands or combs, hence the common name "comb jelly." They are beautifully iridescent in sunlight and usually phosphorescent in darkness. They are important economically because they destroy large numbers of young fish.

Cunner. A rather deep-bodied fish with a long, spiny back fin, found especially about wharf piles and sea walls and sometimes offshore, from Labrador to New Jersey.

Curlew. A large, long-billed bird belonging to the same general group as sandpipers. It ranges in winter to the Pacific Coast of South America, from which it migrates, either by way of the Pacific Coast or by Central America, Florida, and the Atlantic Coast, to the shores of the Arctic Ocean, where it breeds. The long-billed and Eskimo curlews were virtually exterminated during the past century, but fair numbers of the Hudsonian curlew remain.

Cyanea (sī-ā´-nē-ȧ). This is the largest of the Atlantic Coast jellyfishes. In cold northern waters the bell-shaped body may be seven and one half feet across, with tentacles more than a hundred feet long. About ninety-five percent of this great bulk is water. Common sizes are three to four feet across, with thirty- to forty-foot tentacles. Contact with the tentacles produces a severe burning sensation, because of the discharge of hundreds of minute "darts" from the stinging cells. In northern waters Cyanea is red, but the southern form may be a pale bluish or milky white.

Desmid (dĕs´-mĭd). A minute, one-celled, fresh-water alga, often beautifully shaped like a crescent, star, or triangle, and bright green in color.

Diatoms (dī´-ȧ-tŏm). One-celled algae in which the usual green coloring matter is masked by a yellow-brown pigment. The cell walls are impregnated with silica, and after death accumulate in bottom deposits, forming the basis of diatomaceous earth, which is used in polishing powders. Beds of such earth, three hundred feet deep, have been discovered in the Rocky Mountains. Diatoms are the indispensable first links in aquatic food chains, making the nutrient minerals of the water available to the animals that eat them.

Dovekie (dŭv´-kĭ). A maritime bird a little smaller than a robin, belonging to the same family as the auks and puffins. They go ashore only to nest. At sea they are expert divers, and swim under water with their wings, instead of using their feet as the distantly related loons do.

Dowitcher (dou´-ĭch-ĕr). A medium-sized, long-billed shore bird of the sandpiper tribe, seen on the Atlantic Coast during migrations. It winters in Florida, the West Indies, and Brazil, and is believed to nest in northern Canada, east of Hudson Bay.

Dragonfish. In spite of its fierce appearance, only the small inhabitants of the deep sea need fear the dragonfish (also called "viperfish"), for it is only a foot long. It probably spends its entire life in the dark regions that lie more than a thousand feet deep.

Egret, snowy (ē´-grĕt). Often described as the "most dainty and graceful of the herons," the snowy egret was once nearly exterminated because of unrestrained killing for the sake of the beautiful plumes it wears during the breeding season. This bird looks much like a young little-blue heron, but may be distinguished by its yellow feet.

Eider. The eider is a true sea duck, and during its winter migration to the New England and Middle Atlantic Coast spends most of its time offshore, usually over the mussel beds from which it obtains its food by diving. This duck is the principal source of American eider down.

Fathom. A nautical unit of measure equal to six feet.

Fiddler crab. A small, gregarious crab of the beaches and salt marshes. In the male, one of the claws is greatly enlarged into a weapon for defense and attack. Possession of this "fiddle" is in one sense a disadvantage to the male, for it leaves him with only one claw to pick up food, while the female has two. Fiddlers usually live in enormous colonies between the tide lines, each crab in its own small burrow.

Fluke. A name often applied to the summer flounder (Paralichthys dentatus) in the Middle Atlantic and Chesapeake Bay areas. This is one of the more active and predacious flounders, sometimes pursuing schools of fish to the surface. It has a chameleonlike ability to match the color of its background. Average-sized flukes are two feet long.

Foraminifera (fō-răm′-ĭ-nĭf′-ēr-ȧ). A group of one-celled animals usually having limy shells with numerous pores or openings through which long processes of the living substance or protoplasm stream out. The effect is extremely beautiful. After death the shells of these minute creatures sink to the bottom and form chalk beds or deposits of limestone which may be a thousand feet thick. The pyramids of Egypt are built of enormous blocks of limestone formed by fossil Foraminifera.

Frustule. The shell of a diatom, which is in two overlapping parts, like a box and its lid. Being almost pure silica, it is nearly indestructible. The shells are varied in shape and are delicately marked in a great diversity of patterns. These markings are sometimes used to test the power of microscope lenses.

Fulmar (fo͝ol′-mȧr). A bird of the open ocean, belonging to the same family as the petrels and shearwaters. It is a little smaller than a herring gull, spends much of its time on the wing, and is especially active in stormy weather. Its summer range includes Greenland, Davis Strait, and Baffin Bay, while its principal winter resort is off the American coast, especially on the Grand Banks and Georges Banks.

Gannet. On this side of the Atlantic, gannets nest only on rocky cliffs of the Gulf of St. Lawrence, and winter from North Carolina to the Gulf of Mexico. They are large, white birds of the open sea and obtain their food by diving with great force, often from a height of more than a hundred feet. Sometimes a flock of several hundred will attack a school of herring or mackerel.

Ghost crab. A large crab, so pale as to be nearly invisible against the sandy beaches where it lives. It is found from New Jersey to Brazil, and is a common inhabitant of our southern beaches. It is very wary and can outdistance a swift runner. Although it does not hesitate to enter the water when necessary, it lives above the tide line in burrows about three feet deep.

Gill net. A gill net may be anchored on the bottom or buoyed at the top or at almost any intermediate depth, but in any event its position in the water is much like that of a tennis net. Fish are caught in gill nets by thrusting their heads through the meshes and becoming caught by the gill covers, which project slightly, like flaps. A drift gill net is weighted so that it sinks to the bottom and drifts along with the tide.

Gill raker. In breathing, a fish takes in water through the mouth and expels it through the gill openings, which are flanked by the delicate gill filaments that absorb oxygen. The gill rakers are bony projections at the inner entrances to the gill openings. Their function is to strain the food organisms out of the water and also to protect the gill filaments from injury. They have been compared to the human epiglottis, which keeps food from getting into the windpipe.

Glassworm, also called **arrowworm** or **sagitta** (sȧ-jĭt´-ȧ). These are small, elongated, and transparent worms that live only in the sea and are found from the surface to great depths. They are fierce and active predators, and eat large numbers of young fish.

Grebe. Grebes on the water bear a general resemblance to ducks, but if startled will dive rather than fly. They are able to swim considerable distances under water and not uncommonly are caught in fishermen's nets. Usually found in lakes, ponds, bays, and sounds, some grebes venture out to sea for fifty miles or more.

Gyrfalcon (jûr´-fôl-kŭn). A large, predominantly white Arctic falcon that lives chiefly on small birds and lemmings. Occasionally it may wander south in winter to New England, New York, and northern Pennsylvania.

Haddock. A fish of the cod family which lives almost exclusively on the bottom at practically all depths over the continental shelf. The largest haddock on record was thirty-seven inches long and weighed twenty-four and one half pounds.

Hake. Like the haddock, the hakes are members of the cod family, although not at all codlike in appearance, being more slender and tapering fishes. A characteristic feature is the long and feelerlike ventral fin, with which the fish is believed to detect the presence of prey on the bottom.

Hatchetfish. A compressed, silvery, deep-sea fish with highly developed light organs.

Hermit crab. These curious crabs live within the shells of snail-like mollusks, dragging this "house" about with them as protection for their delicate abdomens, which are covered only by a thin skin. When a hermit crab grows too large for its house it must seek a new one, and the inspection of possible quarters is made with great care. Once the selection has been made, the crab whips out of the old shell and into the new with remarkable celerity. Allegedly, it does not confine itself to empty shells, but may forcibly remove the rightful owner.

Holdfast. A rootlike structure, as of algae and other simple plants, for attachment to the substratum.

Hook-eared sculpin. A curious fish with fanlike pectoral fins and conspicuous hooks on the cheeks. It is a cold-water fish, found from Labrador south to Cape Cod and Georges Bank.

Hydroid (hī´-droid). A plantlike animal of the jellyfish group, that is attached at one end and usually has a mouth surrounded by tentacles at the other. The resemblance to a many-branched plant is especially strong when hydroid forms occur in colonies, with a central stalk serving to transport food to the various members.

Jaeger (yā´-gēr). The jaegers belong to the same order of birds as gulls and terns, but in their habits they resemble falcons and other birds of prey. On the high seas, where they winter, they play the part of pirates, forcing gulls, shearwaters, and other birds to give up their booty. During their nesting season on the Arctic tundras, they prey on small birds and lemmings.

Jingle shell. A small mollusk with a very thin shell, usually of a lustrous golden, lemon, or peach color. The empty shells accumulate in windrows on the beach and are said to produce a ringing or jingling sound when disturbed by wind or tide. Jingles are found from the West Indies to Cape Cod.

Killifish. Small minnows of schooling habit, found in droves of thousands of individuals in shallow bays, coves, and marshy places along the coast.

Kittiwake. The kittiwake is a small gull, one of the hardiest of the tribe, for it is truly an oceanic bird and is seldom seen inland except during migrations. This is the gull that follows transatlantic liners for long distances.

Knot. A somewhat robinlike bird of the shore, arriving in the United States from South America in early April. Its nesting grounds were long unknown, but have now been found in the wildest and most remote parts of Grinnell Land, Greenland, and Victoria Land.

Lateral line. The lateral line may be seen in most fishes as a row of pores extending along the flanks from the gill covers to the tail. Internally, these pores communicate with a long, mucus-filled tube which, in turn, is connected with many sensory nerves. The lateral-line organ is believed to allow the fish to detect sound vibrations of a frequency so low that they would scarcely be audible to the human ear. In practice, this means that a fish can sense at a distance the approach of another fish; or can tell that it is coming near to an obstacle, as a wall of rock. According to recent experiments, the lateral line may also help the fish to detect changes in the temperature of the water.

Launce. A slender, round-bodied fish, in appearance something like a small eel. Between the tide lines, it buries itself in the sand while the tide is out. It is a plentiful fish along sandy beaches from Cape Hatteras to Labrador and is abundant over the shoaler parts of the offshore banks. Like most other small, schooling fishes, it forms the food of many ocean predators, including finback whales.

Lemming (lĕm´-ing). A small, mouselike rodent chiefly of Arctic regions, with very short tail, small ears, and furry feet. The Lapland lemmings are remarkable for the mass migrations which they periodically undertake. At such times they advance in great bands in the chosen direction, heedless of obstacles in their path. When they come to the sea they rush into it and are drowned.

Line trawl. Line trawling is an old-fashioned method of fishing for groundfish which has not been wholly superseded by the modern diesel-engined otter trawler. In line trawling, each vessel carries dories, from which the gear is set. A trawl line consists of a long ground line to which short baited lines are attached at intervals of about five feet. Each end of the long line is anchored and marked with a buoy. At intervals, the fishermen take up the lines and remove the fish. Sometimes (in "underrunning" the trawl) the line is merely passed across the dory, the fish removed, the hooks rebaited, and returned at once to the water.

Longspur, lapland. A bird of the finch and sparrow tribe, about the size of a song sparrow. In winter, longspurs are occasionally to be seen in the northern United States and southern Canada, but summer finds them on nesting grounds beyond the tree line in northern Canada and in Greenland and scattered Arctic islands. On the western plains they are described as occurring in "long straggling flocks, all singing together."

Lookdown fish. A very curious fish common from Chesapeake Bay southward. Its body is high and compressed from side to side, and is a beautiful silvery color with opalescent lights. The long, straight profile and high "forehead" give a distinct impression that the fish is looking down its nose.

Marsh samphire (săm´-fīr). The marsh samphire or glasswort is a plant of the salt marshes that turns a vivid red in autumn, forming patches of brilliant color.

Marsh treader. A long and slender-bodied water insect that walks about deliberately over the leaves of water lilies or on the surface film, watching for the mosquito larvae, water boatmen, and small crustaceans on which it preys.

May fly. The greater part of the life of a May fly is spent in the immature stage, when it lives in clean, fresh water as long as three years, burrowing into banks and under stones or running about over the bottom. At maturity it emerges, mates, lays it eggs, and dies, all within one or two days. The life of the adult May fly has become a symbol of a brief and ephemeral existence.

Medusa. The familiar jellyfish in the shape of a bell, umbrella, or disc is known as a medusa. Some jellyfishes have alternating medusa and hydroid stages in their life history. (See hydroid.)

Menhaden (mĕn-hā´-d´n). A schooling fish closely related to shad and herring, found from Nova Scotia to Brazil. It is caught in large quantities for the preparation of oil, meals for stock feeding, and fertilizer, but is not a food fish. It has been described as the prey of every larger predacious animal that swims, including whales, porpoises, tuna, swordfish, pollock, and cod.

Merganser (mẽr-găn´-sẽr). Mergansers are fish-eating ducks that are expert divers and underwater swimmers. The bills are equipped with sharp, toothlike points which are excellently adapted for catching and holding slippery prey.

Mnemiopsis (nē-mĭ-ŏp´-sĭs). This ctenophore reaches a length of four inches and occurs in swarms from Long Island to the Carolinas. It is glitteringly transparent and very phosphorescent.

Moon jelly. (See Aurelia.)

Nereis (nẽr´-ē-ĭs). An active and graceful creature to watch, Nereis is a marine worm that may be from two or three to twelve inches long, depending on the species. It is found under stones and among seaweed in shallow water, and at times swims at the surface. The usual color is bronze, with a beautiful iridescent sheen. Its strong, horny jaws equip it for its life as an active predator.

Noctiluca (nŏk´-tĭ-lū´-kȧ). This single-celled animal (about 3/100 of an inch in diameter) is one of the principal light producers of the sea, at times making large areas glow with an intense phosphorescent light. By day, floating swarms of Noctiluca may tinge the sea with red.

Oarweed. A brown seaweed of the genus Laminaria, all of which are large, with broad, leathery fronds. The larger specimens grow in deep water but often are torn up and washed ashore. Other common names for members of the group are "devil's apron," "sole leather," and "kelp." These algae are among the largest plants known. A related Pacific Coast species may be several hundred feet long.

Old squaw. A sea duck noted for its restless and lively disposition, its noisiness, and its disregard of stormy winter weather. It breeds on the Arctic Coast and winters south to the Chesapeake Bay and the coast of North Carolina. The long tail feathers of the male distinguish it at once from any other duck.

Orca (ôr´-kȧ). The orca, or killer whale, is a member of the dolphin family, but is easily distinguished from its relatives by the very high fin on its back. Packs of orcas travel rapidly at the surface of the sea, attacking whales, dolphins, seals, walruses, and large fishes. They are exceedingly strong and bold. Even large whales appear to be paralyzed with fear at their approach.

Otter trawl. An otter trawl is a large cone-shaped bag of netting which is towed along the ocean's bottom. The average net is about 120 feet long, and 100 feet wide at the mouth. During the towing, the mouth opens to a height of about fifteen feet, being held open by two heavy oak doors, so adjusted that their resistance to the water makes them pull away from each other. The doors, in turn, are attached by long towing lines to the vessel.

Pandion (păn-dī´-ŏn). The scientific name of the osprey.

Petrel, Wilson's (pĕt´-rĕl). These little birds, often called Mother Carey's chickens, visit the coast of the United States during the summer, and in winter return to their nesting grounds on islands off the tip of South America, some within the Antarctic Circle. They are familiar to many as the swallow-like birds that follow in the wake of vessels, apparently dancing on the surface of the water.

Phalarope (făl′-ȧ-rōp). A small bird, between a sparrow and a robin in size. Although it belongs to the shore-bird tribe, its winter range makes it a bird of the open ocean. During migration phalaropes are to be found off our coast in great numbers, but they continue southward, probably well beyond the Equator. They are expert swimmers and feed on plankton when they are at sea. They are said sometimes to alight on the backs of whales to pick off attached sea lice.

Plankton. Derived from a Greek word that means "wanderers," the term plankton is applied collectively to all the minute plants and animals that live at or near the surface of oceans or lakes. Some members of the plankton are wholly passive and drift to and fro with the currents; others are able to swim about actively in search of food. All, however, are subject to the stronger movements of the surface waters. Many sea creatures are temporary members of the plankton during infancy. This is true of most fishes and of bottom-living clams, starfish, crabs, and many other animals.

Pleurobrachia (plo͝or′-ō-brā-kĭ-ȧ). This is a small ctenophore—about half an inch to an inch long—with very long tentacles which may be white or rose-colored. It destroys large numbers of young fish wherever it is abundant.

Plover (plŭ′vẽr). Plovers are shore birds that do not, as a rule, run at the edge of the surf as sandpipers do, but remain higher up on the beach. Among the most familiar kinds are the killdeer and ringneck plovers. As further distinguished from sandpipers, they run about with heads up, then probe suddenly as robins do, instead of constantly probing and dabbing. Plovers nest in Canada and the Arctic (a few species in the United States) and winter as far south as Chile and Argentina.

Portuguese man-of-war. Many people have seen the beautiful blue float of this creature drifting at the surface, especially in tropical waters or in the Gulf Stream. This float acts as an air vessel or sail, and has hanging tentacles that may stretch to a length of forty to fifty feet for anchorage. The Portuguese man-of-war belongs to the same general group as jellyfish and is considered perhaps the most dangerous member of the group, for its sting can cause serious illness or even death.

Pound net. A sort of underwater maze formed of netting attached to stakes driven into the bottom. The opening is so placed that the usual paths of the fish take them into it, and after they have passed through several compartments of the pound it is very difficult to find their way out again. In the last compartment—the "pot" or "crib"—there is also a floor of netting.

Prawn. A shrimp. The two names are often used interchangeably, or "prawn" may be applied to larger specimens, and "shrimp" to smaller.

Ptarmigan (tär´-mĭ-găn). The ptarmigan is a grouselike bird of the Arctic tundras of both the eastern and western hemispheres. In winter, when snow covers the tundra's food supplies, it migrates in immense flocks into protected river valleys of the interior. Occasional specimens have been seen in winter in Maine, New York, and other northern states.

Pteropod (tĕr´-ō-pŏd). A kind of mollusk closely related to the common snail, but bearing little resemblance in appearance or habit to that prosaic creature. Pteropods live in the open waters of the sea, where they swim gracefully through the upper layers. Some have shells of paper thinness; others are without shells and beautifully colored. Sometimes they occur locally in enormous numbers, and are eaten in large quantities by whales.

Purse seine. A purse seine is a net of the encircling type, used in deep water to capture fish that school at the surface. Fish must be visible to be caught in a purse seine—either as dark patches on the water in daylight, or by the phosphorescent glow they stir up on dark nights. The net is dropped into the water in such a way that it hangs in a vertical wall in the shape of a circle, in the center of which is the school of fish. The net is then "pursed" or shirred together by drawing in the lines run through its lower border. The next operation is to take in the slack of the net, concentrate the fish in the "bunt," or section where the twine is strongest, and bail them out with a kind of dip net.

Radiolaria (rā´-dĭ-ō-lā´-rĭ-ȧ). Radiolaria are one-celled animals that live only in the sea and are sometimes large enough to be seen with the unaided eye. Usually they are encased in a skeleton of silica which is exquisitely constructed like a star or a snowflake, with the living substance streaming out through perforations in the skeleton in long, raylike strands. Like the Foraminifera (q.v.) their skeletons sink to the bottom and occur in enormous numbers in marine deposits.

Red clay. A bottom deposit characteristic of the great depths of the ocean (over three miles deep), which carpets a larger area than any other type of deposit. Its basis is hydrated silicate of alumina, and it contains very few organic remains because of the depth at which it lies.

Round-mouthed fish. An oceanic fish that lives at mid-depths and possesses rows of phosphorescent organs with black rims and silver centers. The fish itself may be pale gray to black, depending on the depth at which it lives. (The deeper and darker the water, the darker the fish.) The mouth is extremely large and round when opened; hence the common name.

Rynchops (rĭng´-kŏps). The scientific name of the black skimmer.

Salpa. Salpae or salps are transparent, barrel-shaped animals found in the sea. A single individual is an inch or more long, and many individuals may live together in colonies or chains. This is one of the creatures that show the beginnings of the stiffening rod that is perfected as the backbone in the vertebrates, but it is probably a side branch in the evolutionary tree, which did not lead directly to the development of vertebrates.

Sand bug. Sand bugs are common on beaches from Cape Cod to Florida, where they live in great colonies between the tide lines. When the sand looks strongly pitted after a wave has washed over it, investigation will usually show that there are sand bugs scrambling in the film of water. They are covered with an oval shell, under which the tail or abdomen is bent forward for protection. They are distant cousins of the hermit crab, which resorts to a different device to protect its thin-skinned abdomen (see hermit crab), and are sometimes called "hippa crabs" from their scientific name, Hippa talpoida.

Sand dollar. If all marine animals were as conveniently fashioned as the sand dollar, their identification would be a simple matter. The round, flattened shape of the test or shell accounts at once for its common name, and the star-shaped figure beautifully etched on the shell proclaims its relationship to the starfish. Usually the sand dollar lives on bottoms a little distance from the shore, but it is often washed up on beaches, where its shells are reasonably common objects. In life, the shell is covered by soft, silky spines.

Sand eel. (See Launce.)

Sanderling. Sanderlings are fairly large sandpipers and are among the characteristic birds of the shoreline. They make one of the longest of bird migrations, nesting within the Arctic Circle and wintering as far south as Patagonia.

Sand flea. These small crustaceans are important scavengers of the beaches, promptly devouring dead fishes and all kinds of organic refuse. Turn over a heap of damp seaweed and dozens of beach fleas, usually less than half an inch long, will spring out with great agility. Some forms live in shallow water, others in wet sand or seaweed.

Scallop. The empty shells of scallops are common objects on both east and west coasts. The shells are fan-shaped, with strong radial ridges running from the base of the fan, which also sends out laterally projecting wings in many species. The scallop is an edible mollusk like the oyster and the clam, but only the large, strong muscle that opens and closes the shells is eaten. Only this part of the scallop is seen in markets. Scallops are by no means sedentary shellfish, but swim together through the water with an erratic, darting motion, achieved by rapidly opening and closing the shells.

Scomber (skŏm´-bẽr). The scientific name of the mackerel.

Scup or porgy. This bronze and silvery fish is abundant in the coastal waters from Massachusetts to South Carolina. Some scup make regular migrations from wintering grounds off the Virginia coast to New England, spawning off the coasts of Rhode Island and Massachusetts. Usually they live on the bottom, but sometimes they school at the surface like mackerel.

Sea anemone. A peacefully feeding sea anemone strongly resembles a chrysanthemum, but as soon as it is disturbed, this illusion of flowerlike beauty is dispelled and we see a rather unattractive animal, barrel-shaped and flabby. The "flower petals" are the numerous tentacles which the creature expands in feeding to capture small animals by shooting stinging darts into them. Sea anemones are related to jellyfish and coral animals. They are often delicately and beautifully colored, and range in size from a sixteenth of an inch to several feet across. A few specimens are often to be seen in tide pools, or growing attached to wharf pilings.

Sea cucumber. Sea cucumbers bear scarcely any family resemblance to their relatives, the starfish and sea urchins. They are somewhat wormlike in appearance, with a tough, muscular skin. They move sluggishly over the sea bottom, swallowing sand or mud from which they extract small bits of organic food. They have a strange method of defense when harassed by enemies: they expel their internal organs en masse, later to regenerate them at leisure. Dried sea cucumbers are the "trepang" or bêche-de-mer from which the Chinese make soup, and sea urchins containing eggs are eaten in Europe.

Sea lettuce. A bright green seaweed of flattened, leafy appearance. Although the fronds are tissue-paper thin, this species often grows on rocks exposed to heavy pounding by waves.

Sea raven. This fish is perhaps the most bizarre member of the sculpin tribe, with its large spiny head, ragged fins, and prickly skin. Found in coastal waters from Labrador to the Chesapeake Bay, it is most abundant north of Cape Cod. When lifted from the water it may inflate its body like a balloon, and if thrown back into the water will float helplessly on its back. It is not a market fish, but shore fishermen often save their catches of "ravens" to use for lobster bait.

Sea robin. The sea robin is a fish found chiefly from South Carolina to Cape Cod, with a few living as far north as the Bay of Fundy. In appearance it suggests the sea raven and other sculpins, having a broad head and large pectoral fins (the fins just behind the gills). Often it lies on the bottom with these fanlike fins outspread and will bury itself in the sand up to the eyes if disturbed. The sea robin eats everything from shrimps, squid, and shellfish to small flounders and herring.

Sea squirt. Sea squirts have leathery, saclike bodies, and when touched eject spurts of water from two openings like short teakettle spouts. They grow attached to stones, seaweeds, wharf piles, and the like, straining food animals out of the water by passing it through an elaborate system of internal structures. Sea squirts belong to a group midway between the invertebrates and the true backboned animals. They are eaten in Japan, some South American countries, and in certain Mediterranean ports.

Shearwater. An oceanic bird seen in American coastal waters only when storms occasionally drive it in. One species—the greater shearwater—performs a remarkable migration. Apparently all the members of this species breed on the isolated Tristan da Cunha islands in the South Atlantic ocean. There they nest in deep, grass-lined tunnels in the ground. Every spring they set out on a long northward migration that brings them to the offshore waters of New England, where they remain from mid-May to the middle or end of October. Then they cross the North Atlantic and continue southward off the coasts of Europe and Africa, returning to their island home. It is believed that this circuit of the oceans may take an individual bird two years, and that the breeding cycle may be a biennial one.

Sheepshead. A food fish taken in coastal waters from Massachusetts to Texas. It is nearly always found around old wrecks, breakwaters, and wharfs. The name probably refers to the peculiar shape of the head and more particularly to the large, sheeplike teeth.

Shrimp. A shrimp in life is much like a miniature lobster. Only the jointed and flexible "tail" of the animal is brought into the fish markets, the heads being removed in the packing plants because they contain very little muscle.

Silver eel. An eel in migrating condition is sometimes called a "silver eel" in allusion to the lustrous, silvery color of its underparts.

Silverside. A long, slender little fish with a silvery stripe on its sides, found in either fresh or salt water. Schools of this fish are often abundant off sandy coastlines.

Skua (skū´-ȧ). Skuas are the avian pirates of the high seas. In winter they are fairly numerous on the New England fishing banks, where they terrorize the less belligerent gulls, fulmars, shearwaters, and other birds into giving up the fish, squid, or other food they have caught. The skua nests in Greenland, Iceland, and far northern islands.

Snow bunting. Sometimes called "snowflake," this small bird of the sparrow tribe nests within the Arctic Zone and in winter wanders south as far as southern Canada and the northern United States.

Soldier fly. An insect that gets its name from the gray stripes of the adult. The larvae of some species live in the water as spindle-shaped, dead-looking objects, getting air through a long tube which they push through the surface of the water.

Spadefish. This fish has a body that is almost round and very flat from side to side, and so it is aptly called "moonfish" in some localities. It may be from one to three feet long, and habitually forages about wrecks, pilings, and rocks for encrusting animals. It is found from Massachusetts to South America.

Spot. The fish called spot is so named from a single, round, bronze or yellow spot on each shoulder. It lives in coastal waters from Massachusetts to Texas and is a common food fish. Male spots make a drumming sound like that produced by the croaker, but of less volume.

Squid. The common squid of the Atlantic Coast is about a foot long, and often is to be found in great numbers in coastal waters. Squid are used extensively as bait in fisheries. These animals are noted for their rapid, darting motions and for their ability to change color to match their surroundings. Squid, like oysters and snails, are mollusks, but their shell is reduced to a slender, horny, internal structure called the "pen." These small squid differ in little except size from the almost legendary giant squid, the largest known example of which was fifty feet long, including the extended tentacles.

Sting ray. The flat, roughly quadrangular body of the sting ray and the long, whiplike tail, set with sharp spines, serve to identify it at once. The tail is capable of inflicting an exceedingly painful wound. Sting rays are found along the coast from Cape Cod to Brazil, and occasionally on the shoaler offshore fishing banks. They are closely related to skates and sharks.

Teal. Although small, the blue-winged teal is one of the swiftest of the ducks. Its migratory range extends from Newfoundland and northern Canada as far south as Brazil and Chile, although many of these birds winter in the latitude of the Middle Atlantic States.

Tern. Terns are characteristically birds of the seacoasts. They may be recognized at a glance by their habit of flying about with heads bent to scan the water for signs of fish, which they capture by diving. They nest in enormous colonies on isolated sandy beaches or islands offshore. One species—the Arctic tern—makes one of the longest migrations on record, from the North American Arctic to the Antarctic regions, via Europe and Africa.

Turnstone. A turnstone, once seen, is never forgotten, so startling is the spectacle of this brightly colored black and white and ruddy brown bird of the shore. Its common name refers to its habit of using its short bill to turn over stones, shells, and bits of seaweed in search of sand fleas or other tidbits beneath. It is also called "calico bird."

Water boatman. Almost everyone who has ever stood beside a quiet stream or pond has seen this little insect ferryman sculling across the surface film. The oval boat body is only about a quarter of an inch long; the oars are the hindmost pair of legs, much flattened and fringed with hairs. Surprisingly, some water boatmen fly well, indulging this talent at night, and some produce a kind of music by rubbing the forelegs together.

Whiting. The whiting is a strong and vigorous fish that roves the water from bottom to surface in search of its prey, which consists chiefly of all the smaller schooling fishes. The whiting, sometimes called "silver hake," is closely related to the cod, but is a much more active and slender fish. It is found from the Bahamas to the Grand Banks, and from tidewater down to depths of nearly two thousand feet.

Widgeon grass (wĭj´-ŭn). An aquatic plant which is extensively used as food by waterfowl. Both the small, black seeds and the plant itself are eaten. Widgeon grass grows in brackish water (and sometimes in salt) along the coast, and is also found in interior alkaline waters.

Winged snail. (See Pteropod.)

Yellowlegs. Both the greater and lesser yellowlegs are sometimes called "telltale" or "tattler" from their habit of warning less watchful birds, with loud cries, of approaching danger. The lesser yellowlegs is seldom seen on the Atlantic Coast in spring, for its migration path takes it up the Mississippi flyway to breeding grounds in central Canada. Both species are to be seen on eastern beaches in the fall—large shore birds with rather conspicuous yellow legs. They winter south to Argentina, Chile, and Peru.